NEW
Perennials
for
Canada

Don Williamson

Lone Pine Publishing

Lone Pine Publishing
10145 – 81 Avenue
Edmonton, Alberta T6E 1W9

Website: www.lonepinepublishing.com

Library and Archives Canada Cataloguing in Publication

Williamson, Don, 1962-
   New perennials for Canada / Don Williamson.

Includes index.
ISBN 978-1-55105-842-9

   1. Perennials--Canada. I. Title.

SB434.W34 2010   635.9'6320971   C2009-906961-X

Editorial Director: Nancy Foulds
Project Editor: Wendy Pirk, Kathy van Denderen
Production Manager: Gene Longson
Layout & Production: Rob Tao, Lisa Morley
Cover Design: Gerry Dotto
Cover Photos: front, Daffodil by Margojh/Dreamstime.com; back, Prairie Coneflower and Edelweiss by Don Williamson, and Crocosmia by Tamara Eder

Photo Credits: AASelections 55a; Sandra Bit 10; Rob Broekhuis 154; Joan de Grey 109a, 114, 130, 132b; Tamara Eder 17, 29a, 33a, 60, 61ab, 71, 78, 79a, 90, 91a, 152, 153ab, 186a, 187a, 206, 207a, 227a, 228, 270a, 271bc, 272, 275a, 287; Elliot Engley 37ab, 38a; Erv Evans 155b; Derek Fell 33b, 44a, 55b, 59ab, 62, 63b, 65b, 66, 72, 73b, 75b, 77ab, 85ab, 86, 91b, 93a, 95a, 97, 98a, 99, 101, 103a, 107a, 117, 119b, 123ab, 124, 125b, 126ab, 129a, 135ab, 139b, 141ab, 143b, 144, 145a, 147b, 151, 157a, 159b, 163b, 168, 169, 171b, 176, 177a, 180, 183ab, 191a, 192a, 197ab, 198, 199, 200, 201a, 203a, 208a, 209b, 211b, 215, 217, 218, 219ab, 220, 221a, 222b, 225, 229a, 236, 238a, 240, 241b, 243, 244, 245, 251a, 253a, 254, 255a, 256a, 260, 261ab, 262; Erica Flatt 39ab; Forest Farms 13, 15, 22a, 24a, 26, 27b, 40a, 43b, 45, 76, 81a, 94, 104, 105a, 112, 113, 139a, 142, 146, 147a, 155a, 156, 171a, 172a, 181, 216, 224, 226, 227b, 229b, 233ab, 234b, 235, 248, 257; Saxon Holt 56ab, 57b, 64, 67, 105b, 111, 116, 125a, 127, 134, 137b, 138, 145b, 149a, 157b, 177b, 201b, 210, 241a, 249ab, 250, 251b, 255b, 256b, 259, 263ab; Liz Klose, 174, 275b; Debra Knapke 40b, 115ab, 177b, 185b, 211a; Janet Laughrey 16, 58, 106, 173ab, 175b, 182, 212, 247b; Tim Matheson 21ab, 22b, 23a, 24b, 25ab, 29b, 35a, 41ab, 46ab, 47ab, 51, 68, 69ab, 237ab, 270b, 273, 274, 277, 278; Kim O'Leary 74, 75a, 187b; Allison Penko 23b, 31b, 42, 48, 50b, 92, 93b, 119a, 120, 121ab, 164, 167, 178, 186b, 204, 205b, 208b, 246, 247a, 281; Laura Peters 14, 27a, 28, 30, 32, 34ab, 35b, 36b, 43a, 50a, 63a, 73a, 79b, 80, 81b, 95b, 108b, 110, 128, 129b, 131ab, 132a, 133, 137a, 140, 141a, 149b, 150b, 162, 163a, 165ab, 166ab, 175a, 179ab, 185a, 188a, 189, 190, 193, 194, 195ab, 205a, 207b, 213a, 214, 234a, 282; Photos.com 96, 98b; Robert Ritchie 38b, 107b, 108a, 109b, 271a, 276, 279; Nanette Samol 84, 136, 192b; Mark Turner 65a, 70, 87b, 100, 122, 148, 150a, 172b, 213b, 232, 238b, 239, 242, 258; Don Williamson 1, 3, 4, 18, 19, 20, 31a, 35a, 36a, 44b, 49, 52, 53, 54, 57a, 82, 83, 87a, 88, 89ab, 102, 103b, 118, 143a, 158, 159a, 160, 161, 184, 188b, 191b, 196, 202, 203b, 209a, 221b, 222a, 223, 230, 231ab, 252, 253b, 288; Carol Woo 217d.

We acknowledge the financial support of the Government of Canada through the Book Publishing Industry Development Program (BPIDP) for our publishing activities.

PC: 16

Contents

Acknowledgements

New perennials might be somewhat of a misnomer. Most of these perennials have been around on the planet for a long time, and what is new is your introduction to them. I am grateful for the opportunity to present to you more of the vast array of perennials that will do very well in our northern climate.

I love learning about and experimenting with a variety of plants in my garden and landscape. I am growing or have tried to grow a number of the plants in this guide and would love to be able to grow more of them, but alas, the local climate somewhat limits my selection. That said, there are many beautiful plants in this book that can make your garden unique, no matter where in Canada you live.

Some of you might already own a Lone Pine perennial guide, or one of the other wonderful Lone Pine gardening guides. You might come across a plant in this book that has already appeared in one of the other guides. I included those plants because they did not appear in every guide, and I think Canadian gardeners need to know about and use them more in their gardens.

I thank all my fellow garden writers who I worked with directly or indirectly and who introduced me to a number of perennials that do well in the geographic areas in which these writers live and garden. These people include William Aldrich, Alison Beck, Marianne Binetti, Donna Dawson, Tara Dillard, Don Engebretson, Patricia Hanbidge, Debra Knapke, Laura Peters, Kathy Renwald, Ilene Sternberg and Bob Tanem, and I am truly grateful for their wisdom and knowledge. I thank all the photographers and gardeners that allowed their gardens to be photographed.

I thank Nancy Foulds and Shane Kennedy for the opportunity to do this book, and the rest of the Lone Pine staff for doing an excellent job creating the beautiful and colourful perennial guide you have in your hands.

I thank my sweetheart Darlene for her encouragement and infinite patience while I was working on this book. I think I owe her a few dinners now. I also offer my apologies to my garden, as tending it took a backseat to the writing of this book.

I thank the Creator and Mother Earth for making it possible.

—Don Williamson

The Plants at a Glance

Pictorial Guide in Alphabetical Order

Anise-Hyssop
p. 54

Arum
p. 58

Bear's Breeches
p. 60

Blackberry Lily
p. 62

Bloodroot
p. 64

Bloody Dock
p. 66

Blue Bugloss
p. 68

Blue-Eyed Grass
p. 70

Bluestar
p. 72

Boltonia
p. 74

Bowman's Root
p. 76

Brunnera
p. 78

Burnet
p. 80

Buttercup
p. 82

Calamint
p. 84

Carolina Lupine
p. 86

Cortusa
p. 88

Crocosmia
p. 90

Culver's Root
p. 92

Cyclamen
p. 94

Daffodil
p. 96

Edelweiss
p. 102

Fairy Bells
p. 104

False Indigo
p. 106

Foamy Bells
p. 110

Fringe Cups
p. 112

Gas Plant
p. 114

Germander
p. 116

Globeflower
p. 118

Golden Hakone Grass
p. 120

Green and Gold
p. 122

Hair Grass
p. 124

Hardy Orchids
p. 128

Hellebore
p. 130

Holly Fern
p. 134

Hyssop
p. 136

Ice Plant
p. 138

Indian Pink
p. 140

Inula
p. 142

Irish Moss
p. 144

Ironweed
p. 146

Jack-in-the-Pulpit
p. 148

Jupiter's Beard
p. 152

Kalimeris
p. 154

Kenilworth Ivy
p. 156

Knautia
p. 158

Lady's Slipper Orchid
p. 160

Lemon Balm
p. 162

Lilyturf
p. 164

Liverwort
p. 168

Mayapple
p. 170

Mondo Grass
p. 174

Mountain Mint
p. 176

Northern Maidenhair Fern
p. 178

Ox-Eye Daisy
p. 180

Patrinia
p. 182

Perennial Salvia
p. 184

Perennial Sunflower
p. 190

Pitcher Plant
p. 194

Prairie Coneflower
p. 196

Prairie Poppy Mallow
p. 198

Purple Moor Grass
p. 200

Pussy Toes
p. 202

Ribbon Grass
p. 204

Rodgersia
p. 206

Rosinweed
p. 210

Sedge
p. 212

Self-Heal
p. 216

Sheep's Bit
p. 218

Shooting Star
p. 220

Skullcap
p. 224

Stokes' Aster
p. 226

Stonecress
p. 230

Toad Lily
p. 232

Trillium
p. 236

Tuber Oat Grass
p. 240

Tulip
p. 242

Turtlehead
p. 246

Umbrella Plant
p. 248

Virginia Bluebells
p. 250

Whitlow's Grass
p. 252

Wood Fern
p. 254

Wood Poppy
p. 258

Woodrush
p. 260

Yellow Wax-Bells
p. 262

Perennial gardening is great for people who want a colourful, low- to medium-maintenance landscape with flowers blooming from spring to fall. Canada is an ideal country for perennial gardening, and there is a plethora of perennials to choose from. Our long summer days offer perennials a chance to maximize their growth all season long, and our naturally abundant season of dormancy—winter—provides ample downtime for them to rest up for next year's growth.

Canada's climate is extremely diverse. Tremendous variance in annual temperatures and precipitation occur across the country. Most of us have experienced a 20° C temperature swing in a 24-hour period or have seen a beautiful sunny day turn into a torrential downpour, only to change back into the sunny day again. It is essential that you know what the climate is like where you intend to garden.

Perennials are plants that take three or more years to complete their life cycle. This broad definition usually includes trees and shrubs, but for garden plants, we use the term perennials to refer to herbaceous (nonwoody) plants that live for three or more years. They generally die back to the ground at the end of the growing season and start fresh with new shoots each spring. Some plants grouped with garden perennials do not die back completely; the subshrubs, such as germander, fall into this category. Still others remain green all winter, such as lilyturf.

Gardening in the North

Perhaps the most important question Canadian gardeners ask about a perennial is whether it will grow in their region. Natural Resources Canada's Canadian Forest Service (CFS) scientists have updated the 1967 Agriculture Canada Plant Hardiness Zones of Canada map, which established hardiness zones based on minimum winter temperatures, how long an area is frost-free, maximum temperatures, the amount of summer rainfall, the amount of winter rainfall, the depth and presence of snow cover and maximum wind speed. In a new program called "Going Beyond the Zones," the CFS is calling for experts and the public from across North America and especially Canada to provide information to help develop potential range maps for the species of trees, shrubs and perennials that grow on our continent. Go to http://planthardiness.gc.ca/ for more information about the program and to access the new hardiness zone map.

Know your hardiness zone. If you're not sure, ask someone at your local nursery or a savvy gardener friend. Canada's zones range from zone 0 in the north to zone 8a in a small portion of British Columbia. Proximity to large bodies of water always affects how low temperatures drop in winter. You do not have to go far inland from the oceans or Great Lakes before their warming effects dissipate.

Hardiness zones are only a guide; don't feel intimidated or limited by information on climate zones or by hardiness indicators. Mild or harsh winters, heavy or light snow cover, your soil, fall care and added winter protection, and overall plant health all influence a perennial's ability to live through winter. More than one perennial the experts listed as hardy to zone 5 has turned out to grow very well 100 or even 300 km farther north, in zone 4 or even zone 3.

After zone, soil type is the most important influence on successful plant growth. Soil characteristics vary greatly across the country, and there are hundreds of soil types and distribution patterns. Many of our soils are very good for all kinds of plants, as is evident from the wide variety and high quality of crops, livestock and dairy products produced in here.

Water is another important factor. Many areas in Canada receive ample annual precipitation. If you live in one of those areas, you may not need to provide supplemental watering for your perennials. On the other hand, a number of the plants in this book need moist soil. If you live in a drier climate, and depending on the plants you grow, you may need to provide extra water.

Local topography in your garden may create microclimates—small, sheltered areas out of the wind, for example—that are more favourable for growing plants considered borderline hardy.

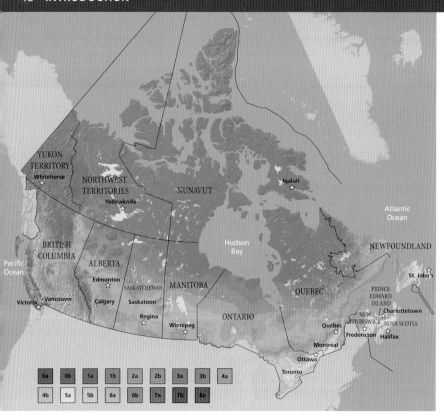

Many enthusiastic and creative people, both amateur and professional, garden in Canada. Growers, breeders, societies, clubs, schools, publications and public gardens all provide information, encouragement and fruitful debate on the subject of gardening. Canadian gardeners take pleasure in their knowledge of planting and propagation methods and have many opinions on what is best for any patch of ground. Outstanding garden shows, public gardens, arboretums and private gardens attract crowds of people. These events and organizations are sources of inspiration as well as valuable information.

Perennials are relatively inexpensive and easy to share with friends and neighbours. The more varieties you try, the more likely you will be to discover what loves to grow in your garden. Get out there and get your hands dirty!

Perennial Gardens

Perennials can be included in any type, size or style of garden. From the riot of colour in a cottage garden to the cool, soothing shades of green in a woodland garden or a welcoming cluster of pots on a front doorstep, perennials open up a world of design possibilities for even the inexperienced gardener.

Perennials can stand alone in a garden or they can be combined with other plants. They form a bridge between the permanent structure provided by woody plants and the season-long, constant colour provided by annuals. Perennials often flower longer and grow to mature size more quickly than shrubs, and most require less care than annuals.

The first step to planning your garden is to decide what type of plants and what garden style you like. Visit garden centres to see the range of plants available, and discuss planting options with the staff. Look around your neighbourhood and community, visit friends and consult books for ideas to use in planning your own garden.

A perennial garden can provide interest throughout the growing season. Selecting perennials that bloom at different times will ensure that some part of your garden is in flower all season. In this book, each perennial description provides blooming seasons.

Colour is often the first thing we notice in a garden. Choose a variety of flower and foliage colours, keeping in mind that different colours have different effects on our senses. Cool colours, such as blue, purple and green, are soothing and make small spaces seem bigger. Warm colours, such as red, orange and yellow, are more stimulating and appear to fill large spaces. White combines well with any colour.

Consider the foliage. Leaves can be bold or dainty, coarse or refined; they can be large or small, light or dark; their colour can vary from any multitude of greens to yellow, grey, blue, purple, silver, white or brown; and they can be striped, splashed, edged, dotted or mottled. Their surfaces can be shiny, fuzzy, silky, rough or smooth. Flowers come and go, but careful attention to foliage will always be interesting.

Umbrella plant

Use a variety of sizes and shapes of plants to make your garden more diverse. The size of your garden bed will influence these decisions, but do not limit a small garden to small perennials or a large garden to large perennials. Use a balanced combination of plant sizes that are in scale with their specific location. (See individual entries and Quick Reference Chart, p. 264.)

Textures can also create a sense of space. Large leaves are considered coarsely textured. Their visibility from a distance makes spaces seem smaller and more shaded. Small leaves, or those that are finely divided, are considered finely textured. They create a sense of greater space and light.

Coarse-textured Perennials
Bear's Breeches
Brunnera
Cortusa
Green and Gold
Inula
Rodgersia
Umbrella Plant

Fine-textured Perennials
Bluestar
Culver's Root
Hair Grass
Holly Fern
Hyssop
Irish Moss
Ribbon Grass
Sedge
Wood Fern

Finally, decide how much time you have to devote to your garden. With good planning and preparation, you can enjoy a perennial garden that is relatively low-maintenance. Try to use plants that perform well with little maintenance and those that are generally pest and disease free.

Low-maintenance Perennials
Bluestar
Ironweed
Kalimeris
Liverwort
Mountain Mint
Ox-eye Daisy
Prairie Poppy Mallow
Pussy Toes
Wood Fern

Bluestar

Getting Started

Once you have some ideas about what you want in your garden, consider the growing conditions. Plants grown in ideal conditions, or conditions as close to ideal as possible, are healthier and less prone to pest and disease problems than plants growing in stressful conditions. Some plants that are considered high maintenance become low maintenance when grown in the right conditions.

Do not try to make your garden match the growing conditions of the plants. Choose plants to match your garden conditions. The levels of light, the type of soil and the amount of exposure in your garden provide guidelines that make plant selection easier. A sketch of your garden drawn on graph paper can help you organize the various considerations as you plan. Knowing your growing conditions can prevent costly mistakes—plan ahead rather than correct later.

Light

Knowing the light conditions in your garden will help you determine where to put your plants. There are four categories of light in a garden: full sun, partial shade, light shade and full shade. Buildings, trees, fences and the position of the sun at different times of the day and year affect the amount of available light.

Full sun locations include south-facing walls with direct sunlight at least six hours per day. Partial shade (partial sun) locations, such as east- or west-facing walls, receive direct sunlight for part of the day (four to six hours) and shade for the rest.

Light shade (dappled shade) locations, such as under a birch tree, receive shade most or all of the day, though some sunlight does filter through to ground level. Full shade locations include the north side of a house or the area under a dense tree canopy, and receive no direct sunlight.

The intensity of full sun can vary. For example, heat that is trapped and magnified between buildings will bake all but the most heat-tolerant plants. A shaded, sheltered hollow that protects your heat-hating plants in the hot, humid summer may become a frost trap in winter, killing tender plants that should otherwise survive.

Inula

Perennials for Full Sun
- Ice Plant
- Jupiter's Beard
- Knautia
- Ox-eye Daisy
- Perennial Sunflower
- Prairie Coneflower
- Prairie Poppy Mallow
- Purple Moor Grass
- Stokes' Aster
- Tulip

Perennials for Full Shade
- Bear's Breeches
- Bloodroot
- Green and Gold
- Holly Fern
- Lady's Slipper Orchid
- Liverwort
- Mayapple
- Northern Maidenhair Fern

Soil

You probably already know if the soil on your property is sandy, heavy clay or predominantly loam, but it is important to know the soil type as well. Taking what you have and amending it with organic matter is always good practice, but first, take a step that is too often ignored by gardeners: have your soil tested. It all starts with the soil.

Private and government laboratories across the country provide soil-testing services for homeowners for relatively little cost. Information and directions on how to take and submit your soil sample are available from the laboratories and online from provincial government sources. The results give you your soil's pH,

Purple moor grass

macro-nutrient content and the percentage of organic material, in addition to directions on how to alter the soil's characteristics to better grow plants. Laboratory soil testing gives more accurate, comprehensive results and better recommendations than the do-it-yourself soil-testing kits available from garden centres.

You can also get a soil food web assay done, which is an analysis of the types and populations of soil microbes and other organisms that make up the soil food web. Plants in their native habitat have a relationship with the soil and the soil life, and they rely on that soil food web to provide them with nutrients they need for growth. Some plants do well in soils with low soil microbe populations, some plants prefer a bacterially dominant soil, some plants like an equal proportion of soil bacteria and fungi, and some plants require a fungally dominant soil. A soil food web assay from a qualified lab can also recommend how to amend your soil to grow the kind of plants you want to grow.

Plants and the soil they grow in have a unique relationship. Many important plant functions go on underground. Soil holds air, water, nutrients, organic matter and a variety of beneficial organisms. Plants depend on these resources, and their roots use the soil as an anchor to hold the rest of the plant upright. In turn, plants influence soil development by breaking down large clods with their roots and by increasing soil fertility when they die and decompose.

Soil is made up of particles of different sizes. Sand particles are the largest. Sand has lots of air space and doesn't compact easily. Water drains quickly out of sandy soil, and nutrients are washed away. Clay particles are the smallest, visible only through a microscope. Water penetrates and drains from clay very slowly. Clay holds the most nutrients, but it compacts easily because there is very little room between the particles for air. Silt particles are smaller than sand particles but larger than clay particles. Most soil is made up of a mixture of all the different particle sizes. These soils are called loams.

Jupiter's beard

If you have nearly impenetrable clay soil, it's probably because of the way the foundation of your home was excavated. When a house is being built and the ground is excavated for the basement, the displaced subsoil is usually just piled nearby. Once the home is built, the pile is levelled off. Topsoil, often only a few centimetres deep, is then spread over top, and you are left to cope with the concrete-like subsoil underneath. To improve this type of soil, work as much organic matter as you can find into the top 30–45 cm of soil. (See "Preparing the Garden," p. 20.)

Perennials for Sandy Soil
False Indigo
Prairie Poppy Mallow
Sheep's Bit
Skullcap
Whitlow's Grass

Whitlow's grass

Perennials for Clay Soil
Bloody Dock
Globeflower
Hair Grass
Prairie Coneflower
Prairie Poppy Mallow
Turtlehead

Particle size influences the drainage properties of your soil, as does slope. Rocky soil on a hillside probably drains very quickly and should be reserved for plants that prefer a very well-drained soil. Low-lying areas retain water longer, and some areas may rarely drain at all. Moist areas suit plants that require a consistent water supply, and areas that stay wet can be used for plants that prefer boggy conditions.

Improve drainage in very wet areas by adding organic matter to the soil, by installing some form of drainage tile or by building raised beds. Never add sand to clay soils—it will make your soil as hard as concrete. Adding copious amounts of organic matter, such as high-quality compost, to clay soil is the best way to turn it into a good growing medium.

Perennials for Moist Soil
Bluestar
Brunnera
Burnet
Cortusa
Crocosmia
Culver's Root
Foamy Bells
Globeflower
Hardy Orchids
Holly Fern
Irish Moss

Mayapple
Northern Maidenhair Fern
Pitcher Plant
Ribbon Grass
Rodgersia
Sedge
Turtlehead
Woodrush

Perennials for Dry Soil
Gas Plant
Germander
Hair Grass
Prairie Poppy Mallow
Pussy Toes
Sheep's Bit

Another aspect of soil to consider is its pH—the measure of acidity or alkalinity. A pH of 7 is neutral; higher numbers (up to 14) indicate alkaline conditions, and lower numbers (down to 0) indicate acidic conditions. Soil pH influences nutrient availability for plants. Although some plants prefer acid or alkaline soils, most perennials grow best in a mid-range pH of between 6.0 and 7.0, and 90% of what we want to grow in Canada prefers a pH of 6.5.

It is possible to make soil more alkaline by adding horticultural lime or to make it more acidic by adding peat moss, sulphur, alfalfa pellets or chopped oak leaves. Note that altering the pH of your soil can take years. If only a few of the plants you are trying to grow require a soil with a different pH from your existing soil, grow them in a container or raised bed where it will be easier to control and amend the pH as needed.

A final soil consideration pertains to plantings along streets. Salt applied to melt the ice in winter accumulates in the soil next to a road. Plants that are not salt tolerant will suffer there. Prairie poppy mallow is a salt-tolerant perennial.

Exposure
Exposure is another important influence in your garden. Some plants are better adapted than others to withstand wind, heat, cold and rain. Buildings, walls, fences, hills, hedges and trees can all influence your garden's exposure.

Plants become dehydrated in windy locations, and strong winds can knock over tall, stiff-stemmed perennials. Plants that do not require staking in a sheltered location may need to be staked in a more exposed one.

Use plants that are recommended for exposed locations, or temper the effect of the wind with a hedge or trees. A solid wall creates wind turbulence on the downwind side, while a looser structure, such as a hedge, breaks up the force of the wind and protects a larger area.

Rodgersia

Perennials for Exposed Locations
- Edelweiss
- False Indigo
- Gas Plant
- Hyssop
- Ice Plant
- Ironweed
- Knautia
- Perennial Sunflower
- Prairie Coneflower
- Prairie Poppy Mallow
- Pussy Toes
- Rosinweed
- Whitlow's Grass

Preparing the Garden

Taking time to properly prepare the area before you plant your perennials will save you time and effort later on. Give your plants a good start by having well-prepared soil that has few weeds and has had organic material added in.

First, loosen the soil with a large garden fork and remove the weeds. Do not work the soil when conditions are very wet or very dry because you will damage the soil structure by breaking down the pockets that hold air and water. Next, amend the soil with organic matter.

All soils benefit from the addition of organic matter because it contributes nutrients and improves soil structure. Organic matter improves heavy clay soils by loosening them and allowing air and water to penetrate. It improves sandy or light soils by binding together the large particles and increasing their ability to retain water, which allows plants to absorb nutrients before they are leached away.

Common organic additives include dried grass clippings, shredded leaves, peat moss, chopped straw, composted manure, alfalfa pellets, mushroom compost and garden compost. Alfalfa pellets supply a range of nutrients including trace elements and a plant growth hormone. Composted cow, chicken and other barnyard manures, available in bags at most garden centres, are wonderful products. Composted horse manure is also an excellent

Edelweiss

Removing weeds & debris

additive and is usually available from stables. If you have access to fresh manure, compost it first, and incorporate it into your beds the season before planting.

Work your organic matter into the soil with a garden fork, shovel, spade or power tiller. If you are adding just one or two plants and do not want to prepare an entire bed, dig holes twice as wide and deep as the rootball of each plant. Add a slow-release organic fertilizer, compost or composted manure to the backfill of soil that you spread around the plant. Fresh chicken or barnyard manure can also be used to improve small areas, but do not put it in the planting hole. Place only a small amount above ground, where it can leach into the soil but not burn the tender roots.

Within a few months, earthworms and other decomposer organisms will break down the organic matter, releasing nutrients for plants. At the same time, the activities of these decomposers will help keep the soil from compacting.

Compost

In forests and meadows, leaves and other plant debris break down on the soil surface, and their nutrients gradually become available to plants. In the garden, we can easily acquire the same nutrient benefits by composting. Compost is a great regular additive for your perennial garden, and good composting methods help reduce pest and disease problems.

Make compost by putting kitchen scraps, grass clippings and fall leaves in a pile, wooden box or purchased composter bin. The process is not complicated: the materials eventually break down if simply left alone. Such "passive," or cool, composting takes a couple of seasons for everything to break down. You can speed up the process and create an "active," or hot, compost pile by following a few simple guidelines.

Use both dry (brown) and fresh (green) materials, with a higher proportion of dry than fresh. Dry matter includes chopped straw, shredded

Wooden compost bins

Ice plant

leaves and sawdust. Fresh matter, which includes kitchen scraps and grass clippings, breaks down quickly and produces nitrogen, which composting organisms use to break down dry matter.

Do not compost meat or dairy products, dog or cat feces, or diseased or pest-ridden material.

Material for compost

Layer the fresh materials evenly throughout the pile between the dry materials. Mix in a shovelful of soil from your garden or previously finished compost to introduce beneficial microorganisms to the compost pile. If the pile seems dry, sprinkle some water between the layers—the compost should be about as wet as a wrung-out sponge.

To speed up decomposition, turn the pile over or poke holes in it with a pitchfork every week or two to get air into it. A well-aerated compost pile reaches temperatures up to 70° C. At this high temperature, weed seeds are destroyed and many damaging soil organisms are killed. Most beneficial organisms, on the other hand, will not be killed unless the temperature exceeds 70° C. You can monitor the temperature near the middle of the compost pile with a thermometer attached to a long probe, similar to a large meat thermometer. Turn your compost when the temperature drops. Turning and aerating the pile stimulates the process to heat up again.

When the material no longer resembles what you put into the pile, and the temperature no longer rises upon turning, your compost is ready to be mixed into your garden beds. Finished compost—which takes as little as a month in warmer areas of Canada—will reward you with organic material rich in nutrients and beneficial organisms. Plus, think of how much less you are sending to the landfill.

If you do not have the space to compost, you can buy bags of it at the garden centre.

Selecting Perennials

Perennials are available as plants or seeds. Purchased plants may begin flowering the same year they are planted; plants started from seed may take two years or more to mature. Starting from seed is more economical if you want large numbers of plants. (See "Propagating Perennials," p. 37.)

Get your perennials from a reputable source, including garden centres, mail-order catalogues, friends and neighbours. Most garden societies promote the exchange of plants and seeds, and many public gardens sell seeds of rare plants. Gardening clubs are also a great source of rare and unusual plants.

Nursery and garden centre staff will answer questions, make recommendations and otherwise assist you. Bring a sketch of your garden space (see "Getting Started," p. 15), with shaded areas, wet areas, windy areas and so forth marked on it. Bring along this helpful book to provide convenient information about the plants as you browse.

Perennials come in two main forms, potted and bare-root. Potted perennials come already growing in pots. Bare-root perennials are pieces of root packed in moist peat moss or sawdust. The roots are typically dormant, though some previous year's growth may be evident, or new growth may be starting. Sometimes the roots appear to have no evident growth, past or present. To make sure you get the best quality potted and bare-root perennials, keep the following points in mind.

Perennials usually come with tags that identify the plant, zone and planting instructions. Poke them into the soil next to the new plants. Next spring, when your perennial bed is nothing but a few stubs of green, the tags will help you locate and identify each plant. A diagram showing the plants' positions in the garden is also helpful if the tags get lost.

Globeflower

Potted plants come in many sizes, and although a larger plant may appear more mature, a smaller one will suffer less from the shock of being transplanted. Most perennials grow quickly once they are put in the garden, so the better buy may well be the smaller plant. Select plants that seem to be a good size for the pot they are in. When you lightly tap a plant out of the pot, you should see roots but they should not wind and twist around the inside of the pot. Healthy roots appear almost white. Avoid potted plants with very dark, spongy roots that pull away with little effort.

The leaves should be a healthy colour. If they are chewed or damaged, check carefully for diseases or insect pests. Do not buy a diseased plant. If you find insects on the plant, don't buy it unless you are willing to cope with them.

Once you get your potted perennials home, plan on planting them as

Ironweed

soon as possible. Until then, water them if they are dry and keep them in a lightly shaded location until you plant them. Remove and discard any damaged growth.

Bare-root plants are most commonly sold through mail order, but

Root mass of root-bound plant

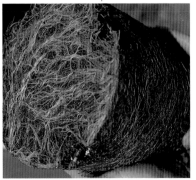

some are available in spring in garden centres. If you are at a garden centre, look for roots that are dormant (without top growth). A bare-root plant that has been trying to grow in the stressful conditions of a plastic bag may have too little energy to recover, or it may take longer to establish once planted.

When you have the bare-root perennials home, cut off any damaged parts of the roots with a very sharp knife or garden scissors. They will dehydrate quickly out of soil, so plant them sooner than you would potted plants. Soak the roots in lukewarm water for one to two hours to rehydrate them (any longer, and you may encourage root or crown rot). Plant the roots either directly in the garden, or into pots with good-quality potting soil until they can be moved to the garden.

Planting Perennials

Once you have planned your garden, prepared the soil and purchased your perennials, it's time to plant.

Potted Perennials

Perennials in pots are convenient because you can arrange them as you like around the bed before you start to dig. To prevent the roots from drying out, do not remove the plants from their pots until immediately before you transplant.

To plant potted perennials, dig a hole about the same width and depth as the pot. Remove the perennial from the pot. If the pot is small enough, hold your hand across the top, letting your fingers straddle the stem of the plant, and then turn it upside down. Never pull on the stem or leaves to get a plant out of a pot. It is better to cut the pot off rather than risk damaging the plant.

If your plants are root-bound, tease apart the roots if they are winding around the rootball. If they have become densely wound around the inside of the pot, cut into or score the root mass with a sharp knife. If there is a solid mat at the bottom of the rootball, remove it; new root tips will only become trapped in the mass. Gently spread out the roots as you plant, teasing a few roots out of the rootball to get them growing outward into the soil.

Place the plant in the prepared hole at the same level it was at in the pot, or a little higher, to allow for the soil to settle. If the plant is too low in the ground, it may rot when water collects around the base. Fill the soil

Loosen rootball before planting (above and below)

in around the roots and firm it down. Water the plant well right away and again every three to four days until it has become established. Water more often if conditions are hot and windy.

Perennials planted in fall may heave out of the ground if they did not have enough time to establish a good root system. To avoid frost heaving, mulch these plants as soon as the ground cools (see p. 29).

The process of cutting into the bottom half of the rootball and spreading the two halves of the mass outward like a pair of wings is called "butterflying." It is an effective way to promote fast growth of pot-bound perennials when transplanting.

Bare-root Perennials

Do not arrange bare-root perennials experimentally around the bed before planting unless you previously put them in temporary pots. Their roots dry out very quickly if left out. If you want to visualize your spacing, poke sticks into the ground or use rocks to represent the plants.

If you have been keeping your bare-root perennials in potting soil, don't be concerned if the roots have not grown enough to knit the soil together and it falls away from the root when you remove the pot. Simply follow the bare-root planting instructions. If the soil does hold together, plant as a potted perennial.

The type of hole you need depends on the root type of your bare-root perennial. For plants with fibrous roots, dig the hole as deep as the longest roots and mound the soil into the centre of the hole up to ground level. Spread the roots out around the mound and cover them with loosened soil.

Stokes' aster

Plants with a taproot need a hole that is narrow and about as deep as the root is long. Use a trowel to open up a suitable hole, then tuck the root into it and fill it in again with the soil around it.

Some plants, such as many irises, have what appear to be taproots, but the shoot seems to be growing off the side of the root rather than upwards from one end. These "roots" are actually modified underground stems called rhizomes. Plant rhizomes horizontally in a shallow hole and lightly cover them with soil so their surface is slightly exposed.

If you find it difficult to distinguish the top from the bottom of some bare-root plants, lay the root in the ground on its side, and the plant will send the roots down and the shoots up.

In most cases, you should try to get the crown of the plant at or just above soil level and loosen the soil that surrounds the planting hole. Keep the roots thoroughly watered until the plants are well established.

Once planted, leave your new perennials, whether potted or bare-root, alone for a while to let them recover from transplant stress. In the first month, you only need to water, weed and watch for pests. Spread organic mulch on the bed around the plants to keep in moisture and control weeds (see p. 29).

If you have prepared your beds properly, you probably won't have to fertilize in the first year. If you do wish to fertilize, wait until your new plants have started healthy new growth, and then apply only a weak organic fertilizer to avoid damaging the sensitive new roots.

Containers

Perennials can also be grown in containers for displays that can be moved around the garden or placed indoors for the winter. Planters can be used on patios or decks, in gardens with very poor soil or in yards where children and dogs might destroy a traditional perennial bed.

Containers are available in a huge variety of sizes, shapes and materials. The important thing is to ensure they have adequate drainage holes in the bottom or sides.

Always use a good quality potting mix intended for containers. Using garden soil is not a good idea; it quickly loses its structure in a container and becomes a solid lump, preventing air, water and roots from penetrating into the soil. Many perennials can grow in the same container without any fresh potting mix for five or six years. Be sure to fertilize and water them more often than those growing in the ground.

When designing a container garden, either keep one type of perennial in each planter and display many different planters together, or mix different perennials in large planters together with annuals and bulbs. The latter choice results in a dynamic bouquet of flowers and foliage. Keep tall, upright perennials, such as perennial salvia, in the centre of the planter; the rounded or bushy types, such as lemon balm, around the sides; and low-growing or draping perennials, such as green and gold, along the edges. Perennials that have long bloom times or attractive foliage work well in planters.

Perennial salvia

Foamy bells

Choose hardy perennials that tolerate difficult conditions. Perennials in planters dry out quickly in hot weather and become waterlogged just as quickly after a few rainy days.

Container perennials are more susceptible to winter damage because the planters provide the roots with very little protection from fluctuations in temperature. The container itself may even crack when exposed to a deep freeze.

It's not difficult to keep planters in great shape through a tough winter. The simplest thing is to move the planter to a sheltered spot. Most perennials require some cold in winter to flower the following year, so find a spot that is cold but not exposed, such as an unheated garage or enclosed porch. Even your garden shed offers plants more protection than they would get outdoors.

If you don't have the space or access to these spots, consider placing

Perennial salvia

your planters in basement window wells where some heat emanates from the windows. Wait until the pots freeze, then layer straw (or Styrofoam insulation if you have mice issues) at the bottom of the well. Set your pots on the straw or insulation, then cover them with more straw or insulation.

You can also winter-proof pots before you plant your perennials. Place a layer of Styrofoam insulation, packing "peanuts" or commercial planter insulation at the bottom of the pot and around the inside of the planter before adding soil and plants. Make sure water can still drain freely from the container. This technique, also useful on high-rise balconies, protects roots from overheating in summer.

As a last resort, you can bury planters in the garden for the winter. Dig a hole deep enough to sink the planter up to its rim and lower the pot into it. It's a messy job, particularly in spring when you dig the planter up, and it's impractical for large planters.

Perennials for Containers

Bloody Dock
Germander
Golden Hakone Grass
Hair Grass
Kenilworth Ivy
Knautia
Lemon Balm
Mondo Grass
Mountain Mint
Perennial Salvia
Ribbon Grass
Stokes' Aster

Caring for Perennials

Weeding

Weeds compete with perennials for light, nutrients and space, and they can also harbour pests and diseases. Many weed seeds, especially those of annual weeds, need light to germinate. A layer of mulch will help prevent weeds from germinating. As your garden matures, some perennials will themselves suppress weeds by blocking the light.

Pull weeds out by hand while they are still small, before they flower and set seed. Or use a hoe to quickly scuff across the soil surface to pull out small weeds and sever larger ones from their roots.

Mulching

Mulch prevents weed seeds from germinating by blocking out the light, and small weeds that do pop up in a mulched bed are very easy to pull. Soil temperatures remain more consistent, and more moisture is retained under a layer of mulch. Mulch also prevents soil erosion during heavy rain or strong winds.

Organic mulches include compost, bark chips, shredded leaves, dried grass clippings and shell and husk materials. These mulches improve the soil and add nutrients as they break down. Shredded newspaper makes wonderful mulch, though it is not as attractive as other mulches. Make sure the leaves and newspaper are shredded because large pieces may keep too much moisture close to the crown of the plant, smothering it.

Spread a 5–7.5 cm layer of mulch over your perennial beds in spring or early summer, depending on the zone you are in. In colder zones, wait until early summer to apply mulch. Gardeners in zone 5 and up can apply mulch in mid- to late spring, which gives the sun time to sterilize the garden topsoil, keeping many fungal diseases under control. Keep the area immediately around the crown or stem of each plant clear. Mulch that is too close to plants can trap moisture and prevent good air circulation, encouraging disease. If the layer of mulch breaks down into the soil over summer, replenish it.

Allow fresh grass clippings to dry for a few days before applying them around your perennials. A layer of fresh grass clippings will heat up before it dries, and this heat can harm the plants.

Applying bark chip mulch

Buy a rain gauge and install it in the ground or at the top of a fence, 6 m away from trees or structures. It's fun to keep track of just how much rain fell during that doozy of a thunderstorm, or to know if your plants are getting enough water.

day, much of the water is lost to evaporation. Moisture left on the plant late in the day may not dry and encourages fungal diseases to develop. Always try to water the ground around the plant, not the plant itself.

Do not overwater. For most perennials, check first that the soil is dry by poking your finger into the top 2.5–5 cm. Some plants need a consistently moist soil; they can be harmed if you allow the soil to dry too much.

Remember that perennials in containers usually need water more frequently than those in the ground. The smaller the container, the more often the plants will need watering. Containers may need to be watered daily during hot, sunny weather. If the soil in your container dries out, water several times to make sure it is absorbed throughout the planting medium. Dig into the soil, and if it is at all dry, water more.

Watering

Once established, many perennials need little supplemental watering if they have been planted in their preferred conditions and are given a moisture-retaining mulch. Planting perennials with similar water requirements together makes watering easier.

Many perennials grow well with an average of 2.5 cm of water per week. Ensure the water penetrates at least 15 cm into the soil. Watering deeply once a week encourages plants to develop a deeper root system. In hot, dry weather, they will be able to seek out the water trapped deep in the ground.

When possible, water in the early morning. If you water in the heat of

Fertilizing

If you prepare your perennial beds well and add new compost to them each spring, you should not need to add extra fertilizer. If you have limited compost, you can mix an organic slow-release fertilizer into the soil around your perennials in spring. Some plants are heavy feeders that need additional supplements throughout the growing season.

Many organic and synthetic fertilizers are available at garden centres. I recommend organic fertilizers for plants growing in the ground; both organic and synthetic fertilizers are acceptable for containers. Most fertilizer instructions recommend a higher rate than is necessary for good plant growth. Never use more than the recommended quantity—it will do more harm than good, and will burn roots. Problems are more likely to be caused by synthetic fertilizers because they are more concentrated than organic types. The addition of any fertilizer can damage the soil microorganism populations, which are mainly responsible for the availability of plant nutrients in the soil.

It is important to support good root development in the first year or two of perennials' growth. Phosphorus promotes root growth, so if you have no compost and the results of your soil test noted that your soil requires phosphorus, apply fertilizers

Knautia

high in phosphorus while your plants are establishing. The typical fertilizer formula is N:P:K (Nitrogen: Phosphorus:Potassium). In the years after your plants establish, nitrogen becomes important for leaf development and potassium for flower and seed development.

Golden hakone grass

Grooming

Many perennials benefit from grooming. Resilient plants, plentiful blooms and compact growth are the signs of a well-groomed garden. Thinning, trimming, disbudding, deadheading and staking are techniques that enhance the beauty of a perennial bed. The methods are simple, but you may have to experiment to get the effect you want.

Thin clump-forming perennials, such as inula and bear's breeches, early in the year when shoots have just emerged. These plants have stems in a dense clump that allow very little air or light into the centre of the plant. Remove half of the shoots when they first emerge to increase air circulation and prevent diseases such as powdery mildew. The increased light encourages compact growth and more flowers. Throughout the growing season, thin any growth that is weak, diseased or growing in the wrong direction.

Pinching or trimming (shearing) perennials is a simple procedure, but timing it correctly and achieving just the right look can be tricky. Early in the year, before the flower buds appear, pinch the plant to encourage new side shoots. Remove the tip and some stems of the plant just above a leaf or pair of leaves. Pinch stem by stem with your fingers, or, if you have a lot of plants, trim off the tops with hedge shears to one-third of the height you expect the plants to reach. The growth that begins to emerge can be trimmed again. Beautiful layered effects can be achieved by staggering the trimming times by a week or two.

Give your plants enough time to set buds and flower. Continual pinching or trimming will encourage very dense growth but also delay flowering. Most spring-flowering plants will not flower if they are pinched back. Pinch early-summer or mid-summer bloomers only once, as early in the season as possible. Pinch late-summer and fall bloomers several times but leave them alone past June. Don't pinch a plant if flower buds have formed—it may not have enough energy or time left in the year to develop a new set of buds. Experiment and keep detailed notes to improve your pinching skills.

Perennial sunflower

Perennials to Pinch Early in the
Season
    Anise-hyssop
    Lemon Balm
    Mountain Mint
    Perennial Sunflower
    Turtlehead

Removing some flower buds to
encourage the remaining ones to
produce larger flowers is called disbudding. It is popular with growers
and gardeners who enter plants in
flower competitions.

Deadheading is removing flowers
once they are finished blooming, and
it serves several purposes. It keeps
plants looking tidy; it prevents the
plant from spreading seeds, and
therefore seedlings, throughout the
garden; it often prolongs blooming;
it spurs root growth; and it helps
prevent pest and disease problems.

Blue-eyed grass

Carolina lupine

Remove flowers that have finished
blooming by hand or snip them off
with hand pruners. Use garden shears
to more aggressively prune back
bushy plants with many tiny flowers,
and plants with a short bloom
period, such as kalimeris, once they
have finished flowering. Shearing
promotes new growth and possibly
more blooms later in the season.

Perennials to Shear Back After
Blooming
    Bloody Dock
    Blue-eyed Grass
    Brunnera
    Carolina Lupine
    Germander
    Jupiter's Beard
    Perennial Sunflower

Removing whole spent flowering stem

Deadheading is not necessary for every plant. Some plants have attractive seed heads that you can leave in place to provide winter interest. Other plants are short-lived, and leaving some of the seed heads in place encourages future generations to replace the old plants. Blackberry lily is one example of a short-lived perennial that reseeds. In some cases, the self-sown seedlings do not possess the attractive features of the parent plant. Deadheading may be required in these cases.

Hellebore

Perennials with Interesting Seed Heads or Fruit

Blackberry Lily
False Indigo
Mayapple
Trillium

Perennials That Self-Seed

Anise-hyssop
Blackberry Lily
Bloodroot
Blue-Eyed Grass
Burnet
Calamint
Fringe Cups
Hellebore
Ironweed
Jupiter's Beard
Knautia
Lemon Balm
Patrinia
Prairie Poppy Mallow
Self-Heal

Staking is the use of poles, branches or wires to hold plants erect. It can often be avoided by astute thinning and trimming, but a few plants always need support to look their best. Three types of stakes are used for the different growth habits that need support. Plants that develop tall spikes, such as inula and blue bugloss, may require each spike to be staked individually. A strong, narrow pole such as a bamboo stick can be pushed into the ground early in the year, and as the spike grows, it can be tied to the stake. Ensure the plant is securely, but not tightly, fastened with soft, plant-friendly ties, or use plant ties from a garden centre.

Use tomato cages or wire hoops called peony rings if staking each stem is not possible. A forked branch can also be used to support single-stemmed plants.

Many plants get top heavy as they grow and tend to flop over once they reach a certain height. The growing conditions for the plant might also be too lush or too shady, causing the plant to be floppy. A peony ring is the most unobtrusive way to hold up such a plant. When the plant is young, push the legs of the peony ring into the ground around the plant. As the plant grows, it will be supported by the ring. The bushy growth also hides the ring.

Other plants, such as knautia, form a floppy tangle of stems. Give these plants a bit of support when young with twiggy branches inserted into the ground; the plants then grow up into the twigs.

Along with thinning and trimming, you can take other steps to reduce the need for staking. First, grow plants in the right conditions. A plant will not do better in a richer soil than is recommended—very rich soil causes many plants to produce weak, leggy growth that is prone to falling over. Similarly, a plant that likes full sun will become stretched out and leggy if grown in shade. Second, use other plants for support. Mix plants that have a stable structure in between plants that need support. The weaker plants may still fall over slightly, but only as far as their neighbours will allow.

Spiral stakes

Burnet

Anise-hyssop

Winter Preparation and Protection

When frost kills your garden in fall, you can either cut down your perennials to the ground (except ornamental grasses and anything that vines or climbs) or leave them to provide winter interest. Instructions for winter care are provided in the individual plant accounts. Keep in mind that if you leave the perennials until spring,

the plants will be floppy and soppy and may be covered in fungus, and you will be unduly stepping in the bed.

Over winter, protect perennials that are borderline hardy in your area by covering the garden with 20–25 cm of straw, hay or bags of leaves once the ground freezes in late fall (pull apart the compacted straw or hay in the bales and fluff it over the area). This fall mulch is particularly important in areas that can't depend on a consistent layer of snow.

Do not mulch before the ground cools, because trapped heat in the soil will prevent the perennials from entering complete winter dormancy; wait until the top 5–7 cm of the ground has frozen. The goal of winter mulch is not to keep the ground warm but to keep it frozen, or, at the very least, a constant, cool temperature until spring.

In late winter or early spring, once the weather starts to warm, pull some of the mulch off to see if the plants have started growing. If they have, pull all of the mulch back, but keep it nearby in case you need to replace it to protect the tender new growth from a late frost. Once you're no longer worried about frost, remove the protective mulch completely.

Lemon balm

Propagating Perennials

Propagating your own perennials is an interesting and challenging aspect of gardening that requires time and space, and can save you money. Seeds, cuttings and division are three common methods of propagating perennials. Each method has advantages and disadvantages.

Starting from Seeds

Starting perennials from seed is a great way to propagate a large number of plants at relatively low cost. Buy seeds or collect them from your perennial garden. Many of the perennials in this book can be started from seed.

Filling cell packs

Prepared seed tray

A cold frame is a wonderful gardening aid used to protect tender plants over winter, to start vegetable seeds early in spring, to harden plants off before moving them to the garden, to protect fall-germinating seedlings and young cuttings or divisions, and to start seeds that need a cold treatment. This mini-greenhouse structure is built so that ground level on the inside of the cold frame is lower than on the outside, so the soil around the outside insulates the plants within. The angled, hinged lid is fitted with glass. The lid lets light in to collect some heat during the day, and it prevents rain from damaging tender new plants. If the interior gets too hot, you raise the lid for ventilation.

Some plants grow well from seed, especially those plants that self-seed easily. Some perennials may only be available as seeds. Some cultivars and varieties don't pass on their desirable traits to their offspring. Other perennials have seeds that take a long time to germinate, if they germinate at all, and may take even longer to grow to flowering size.

Specific propagation information is given for each plant in this book, but a few basic rules apply for starting all seeds. Some seeds can be started directly in the garden (direct sown),

Using folded paper to plant small seeds

but it is easier to control temperature and moisture levels and to provide a sterile environment if you start the seeds indoors. Start seeds in pots or, if you need a lot of plants, flats. Use a sterile soil mix intended for starting seeds. The mix will generally need to be kept moist but not soggy. Most seeds germinate in moderately warm temperatures of about 13–21° C.

Seed-starting supplies available at garden centres include various soil mixes, plastic seed-tray dividers, and heating coils or pads to help keep the soil at a constant temperature.

False indigo

First moisten your soil mix, then fill your pot or seed tray with soil and firm it down slightly—not too firmly or the soil will not drain. Plant only one type of seed in each pot or flat. Plant medium to large seeds individually and spaced out in the pots or trays. If the trays have divided inserts, plant one or two seeds per section. Lightly cover medium-sized seeds with soil. Press large seeds into the soil before lightly covering them. Do not cover seeds that need exposure to light to germinate (the seed packet provides this information).

Sprinkle small seeds onto the soil or fold a sheet of paper in half and place the seeds in the crease. Gently bounce or roll the seeds off the paper in a controlled manner. Some seeds are so tiny that they look like dust. Mix these tiny seeds with a small quantity of very fine sand and spread on the soil surface. Small and tiny seeds may not need to be covered with any more soil.

Water the seeds using a very fine spray if the soil starts to dry out. Try not to disturb the seeds. Place pots inside clear plastic bags to maintain moisture (or use the clear plastic covers that come with the seed trays). Change the bags or turn them inside out when condensation starts to build up and drip. Hold up plastic bags with stakes or wires poked around the edges of the pot. Remove the covers once the seeds germinate.

True leaves look like the mature leaves. The first one or two leaves are actually part of the seed and are called seed leaves, or cotyledons.

The amount and timing of watering is critical. Most germinated seeds and young seedlings will perish if the soil is allowed to dry out. Strive to maintain a consistently moist soil, which may mean watering lightly two to three times a day. As the seedlings get bigger, cut back on the number of times you water, but water a little heavier each time. When the seedlings have their first true leaves, cut back watering to once a day.

Seeds generally do not require a lot of light to germinate, so keep pots or trays in any warm, out-of-the-way place. Once the seeds have germinated, place them in a bright location out of direct sun. Transplant seedlings into individual pots once they have three or four true leaves. Leave plants in plug trays until neighbouring leaves start to touch each other. At this point, the plants will be competing for light and should be transplanted to individual pots.

Do not fertilize young seedlings—the seeds provide all the nutrition they need, and fertilizer causes seedlings to produce soft, spindly growth susceptible to attack by insects and diseases. Use a fertilizer diluted to one-quarter or one-half strength once seedlings have four or five true leaves. Organic fertilizers have a low potential of burning the new, tiny roots.

Many seeds sprout easily as soon as they are planted, and some do not germinate until conditions are favourable. Some seeds bear thick seed coats; some produce chemicals that prevent germination; and some are programmed for staggered germination. In the wild, such strategies improve the chances of survival, but

Seedlings are susceptible to a problem called damping off, which is caused by soil-borne fungi. The afflicted seedling looks like the stem has been pinched at soil level, causing the plant to topple over. The pinched area blackens and the seedling dies. Sterile soil mix, good air circulation, evenly moist soil and a 7 mm layer of peat moss spread over the seedbed all help prevent damping off.

Scarifying seeds with sandpaper

Preparing seeds for cold treatment

Fringe cups

washes away the chemicals that have been preventing germination. Cyclamen is a plant with this type of seeds.

Other thick-coated seeds need to be scratched ("scarified") to allow moisture to penetrate the seed coat and prompt germination. In nature, birds scratch the seeds with gravel in their craws and acid in their stomachs. You can mimic this process by nicking the seeds with a knife or file, or by gently rubbing them between two sheets of sandpaper. Leave the scratched seeds in a dry place for a day or so before planting them to give them a chance to prepare for germination before they are exposed to water. False indigos have seeds that need their thick coats scratched.

Plants from northern climates often have seeds that wait until spring before they germinate. These seeds must be given a cold treatment to mimic winter before they will germinate. Gas plant and kalimeris are examples of these plants. One simple but not very practical method of cold treatment is to plant the seeds in a pot or tray and place them in the refrigerator. Another method is to mix the seeds with some moistened sand, peat moss or sphagnum moss. Place the mix in a sealable sandwich bag and pop it in the refrigerator. With either method, you leave the seeds in the fridge for two months, checking regularly to make sure the planting medium doesn't dry out.

you will have to lower the defences of these types of seeds before they will germinate. For example, soaking thick-coated seeds for a day or two in a glass of water mimics the beginning of the rainy season, which is when it would germinate in its natural environment. The water softens the seed coat and in some cases

As noted in some entries in this book, some plants have seeds that must be planted when freshly ripe. These seeds cannot be stored for long periods of time.

Gas plant

Cuttings

Cuttings are an excellent way to propagate varieties and cultivars that you like but that don't come true from seed or don't produce seed. Each cutting will grow into a reproduction (clone) of the parent plant. Cuttings are taken from the stems of some perennials and the roots or rhizomes of others.

Stem cuttings are generally taken in spring and early summer when plants produce a flush of fresh, new growth, either before or after flowering. Don't take cuttings from plants that are in flower or about to flower: they are busy trying to reproduce. Plants that are busy growing, by contrast, are full of the right hormones to promote quick root growth.

Perennials to Propagate from Stem Cuttings

Bluestar
Germander
Hyssop
Ice Plant
Kenilworth Ivy
Perennial Salvia
Prairie Poppy Mallow
Skullcap
Turtlehead

Because cuttings need to be kept in a warm, humid place to root, they are prone to fungal diseases. Providing proper sanitation (sterile soil mix, clean tools and containers) and encouraging quick rooting will increase the survival rate of your cuttings. Be sure to plant a lot of them to make up for any losses. Dusting with soil sulphur will also help reduce the incidence of fungal disease.

Removing lower leaves

Generally, stem cuttings are more successful and quicker to grow if you include the tip of the stem in the cutting. Some gardeners claim that smaller cuttings are more likely to root and to root more quickly. Others claim that larger cuttings develop more roots and become established more quickly once planted in the garden. Try different sizes to see what works best. Generally, a small cutting is 2.5–5 cm long, and a large cutting is 10–15 cm long.

The size of cuttings is partly determined by the number of leaf nodes on the cutting. The node is where the leaf joins the stem, and it is from

Firming cutting into soil

here that the new roots and leaves will grow. The base of the cutting will be just below a node. You want at least three or four nodes on a cutting. Strip the leaves gently from the first and second nodes and plant them below the soil surface. Retain the leaves on the top part of the cutting above the soil. Some plants have a lot of space between nodes, and others have almost no space at all between nodes. Cut these plants according to the length guidelines, and gently remove the leaves from the lower half of the cutting. Plants with closely spaced nodes often root quickly and abundantly.

Always use a sharp, sterile knife to take cuttings (to sterilize a knife, dip it in denatured alcohol or a 10% bleach solution). Make cuts straight across the stem. Once you have stripped the leaves, dip the end of the cutting into a rooting-hormone powder intended for softwood cuttings. Tap or blow the extra powder off the cutting— cuttings caked with rooting hormone are more likely to rot than to root, and they don't root any faster than those that are lightly dusted.

Plant the cuttings into a sterile soil mix, or keep them in water until they show root growth, and then plant them. If you choose to place them in water, use dark containers instead of clear containers for the best results. The sooner you plant your cuttings or place them in water, the better. The less water the cuttings lose, the less likely they are to wilt and the more quickly they will root.

In pots and trays, use a sterile soil mix intended for seeds or cuttings, or use sterilized sand, perlite, vermiculite or a combination of the three. Firm the soil down and moisten it before you start planting. Poke a hole in the soil with a pencil, tuck the cutting in and gently firm the soil around it. Make sure the lowest leaves do not touch the soil. Space the cuttings far enough apart that adjoining leaves do not touch each other.

Cover the pots or trays with a plastic bag to keep in the humidity. Push stakes or wires into the soil around the edge of the pot so that the plastic will be held off the leaves. Turn the bag inside out when condensation becomes heavy. Poke a few holes in the bag to create some ventilation.

Keep the cuttings in a warm place, about 18–21° C, in bright, indirect light. Use a hand-held mister to gently keep the soil moist without disturbing the cuttings.

Turtlehead

Jack-in-the-pulpit

Most cuttings require one to four weeks to root. After two weeks, give the cutting a gentle tug. You will feel resistance if roots have formed. If the cutting feels as though it can pull out of the soil, gently push it back down and leave it longer. New growth is also a good sign that your cutting has rooted. Some gardeners simply leave the cuttings alone until they can see roots through the holes in the bottoms of the pots. Uncover the cuttings once they have developed roots.

When the cuttings are showing new leaf growth, apply a foliar feed (available at garden centres) using a hand-held mister. Plants quickly absorb nutrients through the leaves, and by foliar feeding, you can avoid stressing the newly formed roots.

Once your cuttings have established, pot them individually. If you rooted several cuttings in one pot or tray, the roots may have tangled together. If gentle pulling doesn't separate them, rinse some of the soil away from the entire clump to free enough roots to allow you to separate the plants.

Pot the young plants in sterile potting soil. Move them into a sheltered area or a cold frame until they are mature enough to fend for themselves in the garden. The plants may need some protection over the first winter. Keep them in the cold frame if they are still in pots, or give them an extra layer of mulch if they have been planted out.

Taking basal cuttings involves removing the new growth from the main clump of a plant and rooting it in the same manner as you would stem cuttings. Many plants send up new shoots or plantlets around their bases. Often, the plantlets will already have a few roots growing. Once separated, these young plants develop quickly and may even grow to flowering size the first summer.

Treat these cuttings in much the same way as you would a stem cutting. Use a sterile knife to cut out the shoot. You may have to cut back some of the shoot's top growth because the tiny developing roots may not be able to support it. You can also let the plant decide what parts are no longer needed. Sterile soil mix and humid conditions are preferred. Pot plants individually, or place them in soft soil in the garden until new growth appears and roots have developed; then you can transplant them where you want them.

Skullcap

Holly fern

Perennials to Start from Basal Cuttings
- Holly Fern
- Jack-in-the-pulpit
- Knautia
- Prairie Poppy Mallow
- Skullcap
- Whitlow's Grass

Root cuttings can be taken from the fleshy roots of perennials that do not propagate well from stem cuttings. Take them in early or mid-spring. The ground is just starting to warm, the roots are about to break dormancy and are full of nutrients the plant stored the previous summer and fall,

and hormones are initiating growth. You may have to wet the soil around the plant so that you can loosen it enough to get to the roots.

You do not want very young or very old roots. Very young roots are usually white and quite soft; very old roots are tough and woody. Use the tan-coloured, fleshy roots.

To prepare your root, cut out the section you will be using with a sterile knife. Cut the root into pieces 2.5–5 cm long. Remove any side roots before planting the sections in pots or planting trays. Plant the roots in a vertical, not horizontal, position and keep them in the same orientation they were when attached to the parent plant. One method of helping you remember which end is up is to cut straight across the tops and diagonally across the bottoms.

Use the same type of soil mix you would for seeds and stem cuttings. Poke the pieces vertically into the soil, leaving a tiny bit of the end poking up out of the soil. Keep the pots or trays in a warm place out of direct sunlight. The root cuttings will send up new shoots once they have rooted, and then they can be planted in the same manner as stem cuttings (see p. 41).

Whitlow's grass

The main difference between starting root cuttings and starting stem cuttings is that the root cuttings must be kept fairly dry; they can rot very easily. Keep the roots slightly moist but not wet while you are rooting them. Do not overwater them as they establish.

Perennials to Propagate from Root Cuttings
Bear's Breeches
Blue Bugloss
Inula
Northern Maidenhair Fern
Virginia Bluebells

Fairy bells

Rhizome cuttings are the easiest means of propagating plants from underground parts. A rhizome sends up new shoots at intervals along its length, and in this way the plant spreads. It is easy to take advantage of this feature.

Dig up a rhizome when the plant is growing vigorously, usually in late spring or early summer. Rhizomes appear to grow in sections. The places where these sections join are called nodes. It is from these nodes that small, stringy feeder roots extend downwards and new plants sprout upwards. You may even see small plantlets already sprouting. Cut your chunk of rhizome into pieces. Each piece should have at least one of these nodes in it.

Fill a pot or planting tray to about 2.5 cm from the top with perlite, vermiculite or seeding soil. Moisten the soil and let the excess water drain away. Lay the rhizome pieces flat on top of the mix and almost cover them with more of the soil mix. If you leave a small bit of the top exposed to the light, the shoots will be encouraged to sprout. The soil does not have to be kept wet; to avoid rot, let your rhizome dry out between waterings.

Once your rhizome cuttings have established, pot them individually and grow in the same manner as stem cuttings (see p. 41).

Stolons are very similar to rhizomes except that they grow horizontally on the soil surface and send roots down at the nodes. Treated them the same way as rhizomes. Pussy toes spreads by stolons.

Perennials to Propagate from Rhizomes
Blackberry Lily
Bloodroot
Blue-eyed Grass
Fairy Bells
Holly Fern

Division

Division is perhaps the easiest way to propagate perennials. As most perennials grow, they form larger and larger clumps. Dividing this clump periodically rejuvenates the plant, keeps its size in check and provides you with more plants. If a plant you really want is expensive, consider buying only one, because within a few years you may have more than you can handle.

How often, or whether, a perennial should be divided varies. Some perennials need dividing almost every year to keep them vigorous. Others are content to be left alone for a long time, though they can be successfully divided for propagation purposes if desired. Still others should never be divided. They may have a single crown from which the plants grow, or they may simply dislike having their roots disturbed. Perennials that do not like to be divided can often be propagated by basal or stem cuttings.

Each entry in this book gives recommendations for division. In general, watch for these signs that indicate a perennial may need dividing:
* the centre of the plant has died out
* the plant no longer flowers as profusely as it did in previous years
* the plant is encroaching on the growing space of other plants sharing the bed.

Begin by planning where you are going to plant your divisions, and have those spots prepared. When those are ready, dig up the entire plant clump and knock any large clods of soil away from the rootball. Split the clump into several pieces with your hands if it is small. Pry a large plant apart with a pair of garden forks inserted back to back into the clump. Square-bladed spades made specifically for slicing through root masses are available. Use a sharp, sterile knife to cut plants with thicker tuberous or rhizomatous roots into sections. In all cases, cut away any dead sections and replant only the newer, more vigorous sections.

Once your clump has been divided into sections, work quickly to get them back into the ground. First work organic matter into the soil and replant one or two of them into the original location. Then plant the

Digging up perennials for division (above & below)

other sections into the prepared spots in the garden, or pot them and give them away.

The larger the sections of the division, the more quickly the plant will grow to blooming size. For example, a perennial divided into four sections will bloom sooner than the same one divided into eight sections. Very small divisions may benefit from being planted in pots until they are more robust.

Newly planted divisions need extra care and attention. Water them thoroughly and keep them well watered. For the first few days after planting, shade them from direct sunlight. A light covering of burlap or damp newspaper is sufficient shelter. Keep divisions that have been planted in pots in a shaded location.

There is some debate about the best time to divide perennials. Some gardeners prefer to divide them while they are dormant, and others believe the plants establish more quickly if divided when they are growing vigorously. Experiment with dividing at different times of the year to see what works best for you. If you do divide perennials while they are growing, you may need to cut back one-third to one-half of the growth to avoid stressing the roots while they are repairing the damage done to them. You can also let the plant decide what foliage it might lose as a result of the division process.

Sometimes if the centre of a perennial dies out, it can be rejuvenated without digging up the whole plant. Dig out the centre of the plant, making sure you remove all of the

Clump of stems, roots & crowns

Cutting apart and dividing tuberous perennials

dead and weak growth. Replace the soil you removed with good garden soil mixed with compost, and sprinkle a small amount of alfalfa pellets on top of the mix. The centre of the plant should fill in quickly.

Perennials That Should Not Be Divided

Carolina Lupine
Gas Plant
Liverwort
Perennial Salvia
Prairie Coneflower
Prairie Poppy Mallow
Shooting Star
Trillium

Lilyturf

Problems & Pests

Perennial beds contain a mixture of different plant species. Because many insects and diseases attack only one species of plant, mixed beds make it difficult for pests and diseases to find their preferred hosts and establish a population. At the same time, because the plants are in the same spot for many years, any problems that do develop can become permanent. Yet, if allowed, beneficial insects, birds and other pest-devouring organisms can also develop permanent populations. Plants selected for this book are generally less susceptible to problems. Included are some native plants that have survived the harsh conditions in Canada for millennia.

For many years, pest control meant eliminating every pest in the landscape. A more moderate approach advocated today is known as IPM (Integrated Pest [or Plant] Management). The goal of IPM is to reduce pest problems to levels of damage that you find acceptable.

You must determine what degree of damage you can live with. Consider whether a pest's damage is localized or covers the entire plant. Will the damage being done kill the plant or is it only affecting the outward appearance? Are there methods of controlling the pest without chemicals? See the Lone Pine books *Lawns for Canada* and the *Garden Bugs* series for Alberta, Ontario and British Columbia for learning about and practising IPM.

IPM includes learning about your plants and the conditions they need for healthy growth, what pests might affect your plants, where and when to look for those pests and how to control them. Keep records of pest damage because your observations can reveal patterns useful in spotting recurring problems and in planning your maintenance regimen.

An effective, responsible pest management program has four steps: cultural controls, physical controls, biological controls and, finally, chemical controls.

Cultural controls are the techniques you use in the day-to-day care of your garden, e.g., growing perennials in the conditions for which they are adapted, and keeping your soil healthy with plenty of organic matter. Choose pest-resistant varieties of perennials. Space your plants so that they have good air circulation and are not stressed by competing for light, nutrients and space. Remove plants that are decimated by the same pests every year. Dispose of diseased foliage. Prevent the spread of disease by keeping your gardening tools clean and by tidying up fallen leaves and dead plant matter at the end of every growing season.

Physical controls are generally used to combat insect and mammal problems, e.g., picking insects off plants by hand is easy if you catch the problem when it is just beginning. Large, slow insects such as Japanese beetles are particularly easy to pick off. Other physical controls include traps, barriers, scarecrows and natural repellents that make a

Prairie coneflower

plant taste or smell bad to pests. Garden centres offer a wide array of options. Physical control of diseases usually involves removing the infected plant or parts of the plant to keep the problem from spreading.

Biological controls use populations of natural predators, e.g., animals such as birds, snakes, frogs, spiders, lady beetles and certain bacteria help keep pest populations at a manageable level. Encourage these creatures to take up permanent residence in your garden. Birds will enjoy a birdbath and birdfeeder in your yard and will feed on a wide variety of insect pests. Many beneficial insects are probably already in your garden, and you can encourage them to stay and multiply by planting appropriate food sources.

Hyssop

Self-heal

Perennials that Attract Beneficial Insects

Anise-hyssop
Blue Bugloss
Calamint
Germander
Hyssop
Inula
Lemon Balm
Mountain Mint
Perennial Sunflower
Prairie Coneflower
Prairie Poppy Mallow
Self-Heal

Plants Not in this Book that Attract Beneficial Insects

Black-eyed Susan
Clover
Daisies
Dandelion
Dill
Fennel
Lavender
Lovage
Queen Anne's Lace
Thyme
Yarrow

Another form of biological control is the naturally occurring soil bacterium *Bacillus thuringiensis* var. *kurstaki,* or *B.t.* for short. It breaks down the gut lining of some insect pests and is available in garden centres. However, *B.t.* can kill many beneficial insects in your garden, so large applications of it are not a good idea. Very small applications of minute amounts may be acceptable.

Chemical controls can do more harm than good and so should only be used as a last resort. These chemical pesticide products can be either organic or synthetic. If you have tried cultural, physical and biological methods and still wish to take further action, find out what pesticides are recommended for particular diseases or insects. Try to use organic options. Organic sprays are no less dangerous than chemical ones, but they will at least break down into harmless compounds, often much sooner than synthetic compounds. See also the environmentally friendly alternatives listed below.

Any pesticide you apply, whether organic or synthetic, disrupts the balance of microorganisms in the soil profile, kills many beneficial insects and initiates the vicious circle of having to use those products to control your pest problems. I would like to see all forms of pesticides used on plants eliminated, or at least severely reduced, and people willing to accept some pest damage.

Pest Control Alternatives

The following common-sense treatments for pests and diseases allow the gardener some measure of control without resorting to harmful chemical fungicides and pesticides.

Ant Control

Mix 750 mL water, 250 mL white sugar and 20 mL liquid boric acid in a pot. Bring just to a boil and remove from the heat. Let cool. Pour small amounts of the cooled liquid into bottle caps or other very small containers and place them around the ant-infested area. You can also try setting out a mixture of equal parts powdered borax and icing sugar (no water).

Baking Soda & Citrus Oil

This mixture treats both leaf spot and powdery mildew. In a spray bottle, mix 20 mL baking soda, 15 mL citrus oil and 4 L water. Spray the foliage lightly, including the undersides. Do not pour or spray this mix directly onto soil.

Baking Soda & Horticultural Oil

This mixture is effective against powdery mildew. In a spray bottle, mix 20 mL baking soda, 15 mL horticultural oil and 4 L water. Spray the foliage lightly, including the undersides. Do not pour or spray this mix directly onto soil.

Blue bugloss

Lady's slipper orchid

Coffee Grounds Spray

Boil 500 g used coffee grounds in 12 L water for about 10 minutes. Cool. Strain out and compost the grounds. Transfer to spray bottle and apply to reduce problems with whiteflies.

Compost Tea

Mix 0.5–1 kg compost in 15 L of water. Let sit for 4–7 days, stirring regularly. Dilute the mix until it resembles weak tea. Use during normal watering or apply as a foliar spray to prevent or treat fungal diseases. Do not spray directly onto plants you intend to eat.

Garlic Spray

This spray is means of controlling aphids, leafhoppers, whiteflies and some fungi and nematodes. Soak 90 mL finely minced garlic in 10 mL mineral oil for at least 24 hours. Add 500 mL of water and 7.5 mL of liquid dish soap. Stir and strain into a glass container for storage. Combine 5–10 mL of this concentrate with 500 mL water to make a spray. Test the spray on a couple of leaves and check after two days for any damage. If no damage, spray infested plants thoroughly, ensuring good coverage of the foliage.

Horticultural Oil

Mix 75 mL horticultural oil per 4 L of water and apply as a spray for a variety of insect and fungal problems.

Insecticidal Soap

Mix 5 mL of mild dish detergent or pure soap (biodegradable options are available) with 4 L of water in a clean spray bottle. Spray the surfaces of insect-infested plants and rinse well within an hour of spraying to avoid foliage discolouration.

NEVER overuse any pesticide. Follow the manufacturer's instructions carefully and apply only the recommended amounts. A large amount of pesticide is not any more effective in controlling pests than the recommended amount. Note that if a particular pest is not listed on the package, the product will not control that pest. Proper and early identification of pests is vital to finding a quick solution.

Shooting star

About this Guide

The perennials in this book are organized alphabetically by their most familiar common names, which in some cases is the proper botanical name. If you are familiar only with the common name for a plant, you will be able to find it easily. The botanical name is always listed (in *italics*). I encourage you to learn these botanical names, because only the true botanical name defines exactly what plant it is everywhere on the planet. Common names are sometimes used for any number of very different plants, and also change from region to region. Sometimes a plant will have more than one common name in one region. Learning and using the botanical names allows you to discuss, research and purchase plants with supreme confidence and satisfaction.

The illustrated "Plants at a Glance" section at the beginning of this book allows you to become familiar with the different flowers quickly, and it will help you find a plant if you're not sure what it's called.

At the beginning of each entry are height and spread ranges, flower colours, blooming times and hardiness zones. At the back of the book, a Quick Reference Chart summarizes different features and requirements of the plants; this chart comes in handy when planning diversity in your garden.

Each entry gives clear instructions and tips for planting and growing the perennial, and it recommends favourite species and varieties. Note: if height or spread ranges or hardiness zones are not given for a recommended plant, assume these values are the same as the ranges at the beginning of the entry. Keep in mind, too, that many more hybrids, cultivars and varieties are often available. Check with your local garden centres when making your selection.

Pests or diseases commonly associated with a perennial, if any, are also listed for each entry. Consult the "Problems & Pests" appendix for information on how to solve these problems.

Anise-Hyssop

Giant Hyssop
Agastache

Height: 60 cm–1.2 m • **Spread:** 46–60 cm • **Flower colour:** purple, white •
Blooms: summer • **Hardiness:** zones 3–8

Bees, butterflies and gardeners love this pretty North American native perennial.
The flowered spikes are attractive, and the sweet licorice scent of the foliage
invites you to touch and smell. Pinching encourages bushy growth, so don't
be afraid to pick a handful of tips and leaves and make yourself a pot of tea to
enjoy in your garden. Both the flowers and leaves are edible and can be used
to garnish your garden salads.

Planting

Seeding: start seeds indoors in late winter, or direct sow after the danger of frost has passed
Transplanting: spring, summer
Spacing: 30–60 cm

Growing

Anise-hyssop grows well in **full sun** and performs adequately in partial shade. This plant will adapt to most soil conditions as long as the soil is **well drained.** Once established, plants are very drought tolerant. Cover plants with a thick layer of mulch in fall for winter protection in the colder zones where snow cover is not guaranteed.

A. foeniculum 'Golden Jubilee'

Anise-hyssop self-seeds but is not invasive. Plants are short-lived, so leave some seedlings in place to replace the older plants as they die out. Divide every two to three years in spring. Pinch the plants back when they are about 15–25 cm tall to encourage dense growth. Deadhead to prolong the blooming period.

Tips

Anise-hyssop works well in mixed beds and borders, wildflower gardens, meadows and when planted with a variety of contrasting ornamental grasses. It is also suitable for an herb or culinary garden where it will attract bees, butterflies, birds and other pollinators.

A. foeniculum

A. foeniculum

A. foeniculum 'Blue Fortune'

Recommended

A. foeniculum is an erect, clump-forming perennial that grows 60–120 cm tall and wide. Tall spikes of fragrant, white or lavender flowers bloom from summer to fall. The grey-green foliage is licorice scented. **'Blue Fortune'** is a sturdy hybrid of *A. rugosa* and *A. foeniculum* that grows to 90 cm tall. It has dark green foliage and bears light purple-blue flowers. **'Golden Jubilee'** is a compact form that grows 51–90 cm tall and 46–60 cm wide. It has aromatic foliage that is bright chartreuse in

A. foeniculum

A. foeniculum cultivar

spring, fading to lime green with age. It produces long-lasting, blue-purple flower spikes from mid-summer to early fall and is hardy to zone 5.

Problems & Pests

Mildew and rust may be a problem during dry spells. Downy mildew, damping off and other fungal problems are infrequent but possible.

The leaves can be used to flavour teas, fruit salads, drinks and to add scent to potpourri and dried flower arrangements.

Arum

Italian Arum
Arum

Height: 20–30 cm • **Spread:** 20–30 cm • **Flower colour:** greenish white •
Blooms: early summer • **Hardiness:** zones 5–8

Arum is an interesting woodland plant that bears its flashy, spotted, spear-shaped foliage from fall through late spring, and as the foliage withers, it produces the flowering stalks. The flowers are not particularly showy, but the plump, red-orange berries that follow are certainly eye-catching. Arum is a plant for our warmer climate zones and needs a good microclimate and some protection in zone 5 to be successful.

Planting

Seeding: in a cold frame in fall
Transplanting: fall
Spacing: 20–30 cm

Growing

Arum grows well in **full sun** or **partial shade**. The soil should be **fertile, humus rich** and **moist**. Keep in mind that this plant sprouts in fall when deciduous trees are losing their leaves and goes dormant just as they leaf out in spring, so a location you consider shaded may be appropriate if it is sunny at the right time of the year. Divide in summer once the plant has finished flowering. Remove the pulp from the seeds before you plant them. Be sure to wear gloves when handling the plant or its seeds because the sap and pulp can irritate the skin and cause stomach upset if ingested.

Tips

Arums are well suited to open woodland gardens and make excellent ground covering companions for shrubs. They are often included in pondside plantings where their attractive green foliage provides interest from spring through fall, when most other plants are dormant.

Recommended

A. italicum produces arrow-shaped or sometimes spear-shaped green, white-veined or marbled leaves. It grows 20–30 cm tall with an equal spread. Leaves emerge in fall and die back in late spring. Greenish white flower spathes are produced in early summer, followed by colourful spikes of bright reddish orange berries.

A. italicum

Subsp. 'Marmoratum' ('Pictum') has distinctive white marbling on the leaves. **'White Wonder'** has very prominent marbling in the leaves.

Problems & Pests

This plant rarely suffers from problems.

The marbled leaves of A. italicum subsp. 'Marmoratum' are a popular addition to fresh flower arrangements.

A. italicum 'Marmoratum'

Bear's Breeches

Acanthus

Height: 90 cm–1.5 m • **Spread:** 90 cm–1.2 m • **Flower colour:** purple, rose purple, white • **Blooms:** late spring to mid-summer • **Hardiness:** zones 5–8

Bear's breeches are great tropical-looking plants, but they can be invasive. The plants spread by rhizomes, and even a small piece left in the ground may start a new plant. Provide natural or artificial barriers to a minimum depth of 20 cm, or plant bear's breeches where they have plenty of space to grow. Great companions are other large-leaved and tropical-looking plants.

Planting
Seeding: start seeds in containers in spring
Transplanting: spring or fall
Spacing: 90 cm

Growing
Bear's breeches will grow in **full sun to full shade** in just about any **well-drained** soil. They prefer a **rich, moist** soil, but a slightly poorer soil keeps them from being too aggressive. They are drought tolerant but do best if given an occasional soaking. Bear's breeches dislike overly humid conditions. These plants need winter protection in the colder zones.

A. spinosus (both photos)

Bear's breeches may require frequent division. Divisions should be made from late fall to early spring. These plants grow best from root cuttings.

Tips
Bear's breeches are bold, dramatic plants that form large clumps. They work well as the central planting in an island bed or at the back of a border. The foliage is almost as striking as the flower stalks.

Recommended
A. hungaricus (Balkan bear's breeches) is more compact than the other listed species. It has deep green, shiny foliage and bears light pink flowers with purple bracts in mid- to late summer.

A. mollis (common bear's breeches) has shiny, dark green foliage. It is less spiny than *A. spinosus.* Tall spikes of white-and-purple bicoloured flowers are borne in late spring to summer.

A. spinosus (spiny bear's breeches) has silvery, very spiny foliage. This species is more tolerant of humid conditions and winter cold and is less invasive than *A. mollis.* The white flowers with purple to rose-purple bracts are borne from early spring to mid-summer.

Problems & Pests
Snails, slugs, powdery mildew and leaf spot can be troublesome. Good ventilation prevents fungal problems.

Blackberry Lily

Belamcanda

Height: 60–90 cm • **Spread:** 30–46 cm • **Flower colour:** bright yellow to orange • **Blooms:** early summer • **Hardiness:** zones 5–8

Blackberry lily is an excellent plant with wonderful foliage, attractive flowers and showy black berries. It looks like an iris, with beautiful foliage suitable for any garden. Blackberry lily thrives with plenty of sun, good drainage and little care and is worth trying in the colder zones if you have a suitable micro-climate. Don't worry that this wonderful plant is short-lived. It self-seeds enough to keep both you and your friends in plants.

Planting

Seeding: sow seeds directly outdoors in fall because a cold period is required for germination
Transplanting: spring or fall
Spacing: 30–90 cm

Growing

Blackberry lily grows well in **full sun** or **partial shade,** in **moist, well-drained** soil of **average fertility** with lots of **organic matter** mixed in. It can adapt to sandy soils and some clay soils, as long as plenty of organic matter has been added. Plants grown in rich soil may need staking.

B. chinensis 'Hello Yellow'

Blackberry lily can be propagated by dividing the rhizomes in spring, or it can be easily grown from seed.

Tips

Blackberry lily is used to provide a vertical element in mixed beds and borders.

Recommended

B. chinensis forms clumps of sword-shaped, green foliage arranged in fans like those of iris plants. It spreads slowly from thick rhizomes. Clusters of star-shaped, yellow to orange flowers with red or maroon spots rise above the foliage in early summer. The individual flowers last only one day but are produced in succession over a long period. In fall, the fruit splits open to reveal shiny black berries.

Problems & Pests

Blackberry lily is mostly pest free. It may suffer from crown rot in wet soils. The foliage may scorch if the soil is allowed to dry out too much.

B. chinensis 'Mixed Colors'

Bloodroot

Puccoon, Indian Paint
Sanguinaria

Height: 15–25 cm • **Spread:** 30 cm • **Flower colour:** white with yellow centre • **Blooms:** late winter to early spring • **Hardiness:** zones 3–8

Bloodroot is one of the first flowers to bloom in native woodlands, and it's the same in your woodland garden. The white blossoms poke out from the leaf litter before the foliage appears. Bloodroot and tree roots co-exist well with little competition for resources. Hot, dry conditions will cause the foliage to disappear before fall, so be sure to enjoy this woodland beauty in spring. It is best to plant these enchanting beauties in groups of three to five or more for a good show while the rhizomes become established.

Planting

Seeding: start seeds in a cold frame in fall; the species will self-seed in the garden to form large colonies
Transplanting: spring
Spacing: 20–30 cm

Growing

Bloodroot prefers **partial to full shade** and **humus-rich, well-drained, moist, fertile** soil. The delicate flowers easily lose their petals, so make sure to provide shelter from the elements, especially the wind. Divide the rhizomes after flowering has finished.

Tips

Bloodroot does best in a shaded, sheltered woodland garden, and it may also be used effectively in a shady rock garden.

Do not ingest any part of this plant. It is **toxic**.

Recommended

S. canadensis is a woodland plant that grows from thickened, branched rhizomes. Each growing tip of a rhizome produces a single leaf and flower. The large, variable, lobed foliage is a dull blue-green to grey-green. The flower has white petals and yellow stamens and blooms in late winter to early spring as the leaf unfolds. **'Multiplex'** ('Flore Pleno') is a sterile selection that bears double flowers. The flowers are much sturdier and last longer than those of the species, and they are less affected by the elements.

S. canadensis

Problems & Pests

Bloodroot may be afflicted with various types of leaf spot.

S. canadensis 'Multiplex'

Bloody Dock

Red-Veined Dock, Bloody Sorrel
Rumex

Height: 20–38 cm • **Spread:** 20–30 cm • **Flower colour:** green, fading to reddish brown • **Blooms:** early to mid-summer • **Hardiness:** zones 4–8

Bloody dock is a great, easy-to-grow perennial with unusual foliage that has a wide variety of uses. It grows quickly, which makes it a good plant for containers, but some gardeners have complained that it is invasive. It does self-sow freely, but vigilant deadheading will stop the plant from showing up in unwanted places in your landscape.

Planting
Seeding: sow seeds in situ or in containers in spring
Transplanting: spring
Spacing: 20–30 cm

Growing
Bloody dock grows well in **full sun to partial shade** in **moist, well-drained** soil of **average fertility**. It is tolerant of heavy clay soil. Plants with shabby-looking foliage can be cut back after flowering. Divide in spring.

Touching the foliage may irritate your skin, and eating too much of the foliage may cause mild stomach upset.

Tips
Bloody dock can be used as an accent or specimen plant. It looks great when mass planted or as an edging plant in beds, borders and around ponds. It can also be used in woodland gardens and in kitchen gardens.

Recommended
R. sanguineus forms clumps of mid- to dark green foliage with deep red leaf veins. The flower stems rise above the foliage in early to mid-summer, reaching 60–90 cm tall, bearing very small, star-shaped flowers that begin green and fade to reddish brown. The flowers are not showy.

Problems & Pests
Plants may be attacked by aphids, slugs and snails and may experience rust, smut and fungal leaf spots.

R. sanguineus subsp. *sanguineus*

The young leaves of bloody dock can be eaten in moderation either raw or cooked. Older leaves become bitter because of the high amount of oxalic acid that develops, but cooking the leaves helps reduce their bitterness. The bitter taste of the older leaves is also unpalatable to deer.

Blue Bugloss

Italian Alkanet
Anchusa

Height: 30 cm–1.5 m • **Spread:** 30–60 cm • **Flower colour:** blue •
Blooms: late spring to mid-summer • **Hardiness:** zones 3–8

Here's a tall beauty that loves the Canadian climate. These plants are short-lived perennials, but if you allow them to go to seed, you'll continue to have plants to enjoy for many years. Blue bugloss is attractive to bees and butter-flies, but deer and rabbits leave them alone.

Planting

Seeding: sow seeds in a cold frame in spring
Transplanting: after the risk of frost has passed
Spacing: 46–60 cm

Growing

Blue bugloss grows best in **full sun** in **moist, fertile, well-drained** soil. It will adapt to a variety of soils, as long as the soil is well drained. Established plants can tolerate short periods of drought. Use mulch to help keep the rootzone moist. The flowering stems can be quite tall, so choose an area sheltered from the wind. Staking may also be required.

Propagate by taking root cuttings in early spring. Deadheading encourages further blooming.

Tips

Blue bugloss is a great plant for the back of the border, providing height and colour early in the season. Blue bugloss also makes a wonderful specimen or accent plant and creates a sea of blue when mass planted.

Recommended

A. azurea forms mounds of narrow, green foliage with upright, erect stems that bear loose clusters of bright blue flowers from late spring to mid-summer. It grows 90 cm–1.5 m tall and 46–60 cm wide. The stems and leaves have stiff hairs. '**Loddon Royalist**' grows 30–90 cm tall and 30–46 cm wide and bears large, very deep blue blooms up to 5 cm across.

A. azurea (both photos)

Problems & Pests

Problems are rare. Plants may be attacked by rust, powdery mildew, cutworms and grubs. Crown rot can occur in poorly drained soil.

The flowering stems can be cut and used in bouquets and fresh arrangements.

Blue-Eyed Grass

Sisyrinchium

Height: 10–51 cm • **Spread:** 15–46 cm • **Flower colour:** blue, violet-blue, purple, pale yellow • **Blooms:** late spring to early summer • **Hardiness:** zones 3–8

These dainty, easy-to-grow plants are from the iris family and are not grasses at all. All species form clumps of grass-like to sword-shaped foliage and bear clusters of star-shaped flowers with yellow throats. They self-seed where they are happy, but it is not usually excessive.

Planting

Seeding: sow seeds in situ in early fall, or start seeds in containers in a cold frame in spring
Transplanting: spring, after the risk of frost has passed
Spacing: 15–30 cm

Growing

Blue-eyed grasses grow best in **full sun** in **moist, well-drained, neutral to mildly alkaline** soil of **average to poor fertility**. Plants are tolerant of partial shade, mild drought and a variety of soil types. Plants can be cut back after flowering to mitigate self-seeding and to refresh the foliage. Divide rhizomes in early spring.

Tips

Blue-eyed grasses have many uses and look great as specimens or mass planted in any bed or border. Use them at the edge of woodland gardens, in native and wilderness gardens, in rock gardens, around ponds and water features, in bog gardens, as edging and in containers.

Recommended

S. angustifolium (pointed blue-eyed grass) has erect green foliage and grows 30–51 cm tall and 15–30 cm wide. It bears deep blue to violet-blue flowers on branched stems in summer. (Zones 4–8)

S. bellum (*S. idahoense* var. *bellum;* western blue-eyed grass) has stiff, upright, deep green foliage and grows 10–30 cm tall and 15–46 cm wide. It produces deep violet-blue to dark purple flowers from late spring to early summer.

S. montanum (strict blue-eyed grass, common blue-eyed grass) grows 15–30 cm tall and 20–30 cm wide and bears violet-blue to purple flowers from late spring to early summer. The grass-like foliage is green.

S. 'Raspberry' has deep green to blue-green foliage and grows 15–30 cm tall and wide. It bears creamy, pale yellow flowers with deep purple-pink veins from late spring to early summer. (Zones 5–8)

Problems & Pests

Blue-eyed grass has no serious problems but may experience aphids, mites and rust.

S. angustifolium

Bluestar

Amsonia

Height: 38 cm–1.2 m • **Spread:** 60 cm–1.5 m • **Flower colour:** shades of light blue, white • **Blooms:** mid- to late spring • **Hardiness:** zones 3–8

The feathery, fern-like foliage of bluestar contrasts beautifully with coarser-foliaged plants. This plant also boasts multi-seasonal appeal, with its spring flowers, excellent summer foliage and magnificent fall colour. Bluestar looks great either integrated at the back of a perennial bed or by itself as a specimen plant. *A. hubrichtii* and 'WFF Select' are worth trying in zone 4 with winter protection.

Planting

Seeding: not recommended
Transplanting: spring
Spacing: 90 cm–1.2 m

Growing

Plant in **full sun to partial shade** in **well-drained** soil of **moderate fertility**. Too rich a soil will result in thin, open growth and not as many flowers. Light trimming after flowering may bring on a second flush of blooms. Bluestar can be propagated from cuttings in spring or from divisions on established clumps in fall.

Tips

The willow-like foliage of bluestar turns an attractive yellow in fall, and its love of moist soil makes it a beautiful addition to the side of a stream or pond as well as in a border.

Be sure to wash your hands thoroughly after handling the plants because some people find the sap irritates their skin.

Recommended

A. hubrichtii (Arkansas bluestar) grows 90 cm–1.2 m tall and 1.2 m wide. It produces light blue to white, star-shaped blooms in spring. This plant prefers a moist soil. The narrow, feathery foliage turns golden yellow in fall. (Zones 5–9)

A. tabernaemontana (willow bluestar) prefers sun but does well in partial shade. It grows 60–90 cm tall and up to 1.5 m wide, producing small lavender blue flowers in May and June. **Var.** *salicifolia* has narrower

A. tabernaemontana with Rudbeckia 'Goldsturm' in front

foliage and more open clusters of flowers than the species. (Zones 3–9)

A. **'WFF Select'** was discovered and introduced by White Flower Farm in Connecticut. It grows 38 cm tall and spreads 60–90 cm wide and bears deep lavender blue flowers and solid, dark green foliage. This hybrid may be listed as *A.* 'Blue Ice' in some catalogues. (Zones 5–9)

Problems & Pests

Bluestar may have occasional problems with rust.

A. hubrichtii

Boltonia

Boltonia

Height: 90 cm–1.8 m • **Spread:** up to 1.2 m • **Flower colour:** white, mauve or pink, with yellow centres • **Blooms:** late summer and fall • **Hardiness:** zones 4–8

There are plenty of reasons to grow this outstanding perennial. Boltonia is tall, easy to grow, pest and disease free and it blooms profusely for four weeks or more! It is great for providing fresh colour to the garden late in the season. Sometimes called thousand-flower aster, the flowers are tiny, but when one stalk produces hundreds of them, it doesn't really matter. 'Pink Beauty' supplies fresh colour in August, and 'Snowbank' bursts forth in September.

Planting
Seeding: start seeds in a cold frame in fall
Transplanting: spring or fall
Spacing: 90 cm

Growing
Boltonia prefers **full sun** but tolerates partial shade. It prefers soil that is **fertile, humus rich, moist** and **well drained** but adapts to less fertile soils and tolerates some drought. If the plants grow too tall for your liking, cut the stems back by one-third in June. Divide in fall or early spring when the clump is becoming overgrown or seems to be dying out in the middle.

B. asteroides

The stout stems rarely require staking. If necessary, circle the plant with a peony hoop or install twiggy branches while the plant is young. The growing branches will hide the hoop or twiggy branches.

Tips
This large plant can be used in the middle or at the back of a mixed border, in a naturalized or cottage-style garden or near a pond or other water feature. The small, narrow foliage is not particularly showy. Plant boltonia where it will not be noticed until it blooms.

A good alternative to Michaelmas daisy, boltonia is less susceptible to powdery mildew.

Recommended
B. asteroides is a large, upright perennial with narrow, greyish green leaves. It bears lots of white or slightly purple daisy-like flowers with yellow centres. **'Pink Beauty'** has a looser habit and bears pale pink flowers. **'Snowbank'** has a denser, more compact habit and bears more plentiful white flowers than the species.

Problems & Pests
Boltonia has rare problems with rust, leaf spot and powdery mildew.

B. asteroides 'Snowbank'

Bowman's Root

Indian Physic
Porteranthus

Height: 60–90 cm • **Spread:** 30–60 cm • **Flower colour:** white, pink-white
Blooms: late spring to mid-summer • **Hardiness:** zones 4–8

This rare, attractive plant is native to Ontario and the eastern United States. Bowman's root is quite tolerant of competition from tree roots and thrives in moist, deciduous woodlands. The star-shaped flowers have narrow petals that dance and flutter in a light breeze. Bowman's root is well worth trying in zone 3.

Planting

Seeding: sow seeds in containers in a cold frame, or direct sow in spring or fall
Transplanting: late spring
Spacing: 30–60 cm

Growing

Bowman's root grows best in **partial to light shade**. It grows fairly well in full shade and can take full sun if it receives shelter from the hot afternoon sun. The soil should be **moist, well drained, slightly acidic** and **humus rich,** though this plant tolerates a range of soils. The flowering stems may require support. Divide in spring or fall.

Tips

Bowman's root looks best when planted en masse. Use it at the edge of a woodland garden, in beds and borders, in containers and as a cut flower.

Recommended

P. trifoliatus (*Gillenia trifoliata*) forms erect mounds of dark green, compound foliage that turn red to reddish bronze in fall. Each leaf has three leaflets with toothed margins. Airy clusters of white and sometime pink-flushed, slightly nodding flowers bloom from late spring to midsummer and sit atop wiry, red stems. Reddish sepals subtend the flowers and persist after the petals drop. The somewhat hairy seedpods can be left on the plants for winter. **'Pink Profusion'** bears pink and white bicoloured flowers that are mostly pink. New foliage is dark reddish purple to reddish bronze and turns dark green with age.

P. trifoliatus (both photos)

Problems & Pests

Plants are susceptible to rust, and slugs find young plants tasty.

The dried root bark of Bowman's root was used to treat a number of different ailments, including colds, diarrhea, constipation, indigestion, breathing problems, insect stings and rheumatism. It was also used as an emetic— "emetic" being the polite term for stuff that makes you vomit.

Brunnera

Siberian Bugloss
Brunnera

Height: 30–46 cm • **Spread:** 46–60 cm • **Flower colour:** blue • **Blooms:** spring • **Hardiness:** zones 3–8

Wondering what to grow in your shady garden instead of hosta? Many wonderful shade plants are perfectly suited to our northern climate, and brunnera happens to be one of the best. The richly variegated foliage of 'Dawson's White' and 'Hadspen Cream' really brighten the floor of a woodland garden, and the clear blue flowers only improve the show. Brunnera combines effortlessly with many different shade plants and happily grows undisturbed for many years.

Planting

Seeding: start seeds in a cold frame in early fall or indoors in early spring
Transplanting: spring
Spacing: 30–46 cm

Growing

Brunnera prefers **light shade** but tolerates morning sun with consistent moisture. The soil should be of **average fertility, humus rich, moist** and **well drained**. The species and its cultivars do not tolerate drought. Divide in spring when the centre of the clump appears to be dying out.

Cut back faded foliage mid-season to produce a flush of new growth.

Tips

Brunnera makes a great addition to a woodland or shaded garden. Its low, bushy habit makes it useful as a groundcover or as an addition to a shaded border.

Recommended

B. macrophylla forms a mound of soft, heart-shaped leaves and produces loose clusters of blue flowers

B. macrophylla 'Dawson's White'

all spring. **'Dawson's White'** ('Variegata') has large leaves with irregular creamy patches. **'Hadspen Cream'** has leaves with creamy margins. Grow variegated plants in light or full shade to avoid scorched leaves. **'Langtrees'** has blue flowers and large leaves with silver spots.

B. macrophylla

Burnet

Sanguisorba

Height: 60–90 cm • **Spread:** 60–90 cm • **Flower colour:** white, deep maroon, deep purple-red • **Blooms:** early summer to early fall • **Hardiness:** zones 2–8

Got a spot in your landscape that is consistently moist and turns into a mud bog after a rain? Consider using burnets in a drainage garden. In nature, burnet is found in swampy, boggy areas and moist meadows, and it is a good choice for planting in such areas with other moisture-loving plants. Burnets form mounds of attractive, mostly basal, compound foliage with bottlebrush-like spikes of tiny flowers that rise above the mounds. Each flower has four sepals and four long, showy stamens but no petals. They spread by rhizomes, so they have the potential to spread to other areas of your garden.

Planting

Seeding: start seeds in a cold frame in fall or spring, or direct sow in fall or spring

Transplanting: spring, after the risk of frost has passed

Spacing: 30–90 cm

Growing

Burnets grow best in **full sun** in **consistently moist, well-drained** soil. The plants tolerate partial shade. Taller plants may require staking. Divide in early spring. Shabby foliage can be cut back after flowering. Deadhead flowers to prevent unwanted self-seeding.

Tips

Burnet can be used in bog gardens, wildflower meadows and around water gardens and ponds. It looks good as an accent plant or when mass planted and makes a good cut flower. *S. officinalis* also does well in a kitchen or herb garden.

Recommended

S. canadensis (Canada burnet) has large, dense spikes of white flowers on branched flower stems, blooming from early summer to early fall. The foliage is silver-grey to grey-green. Plants grow 1.8 m tall and spread 60–90 cm and prefer an acidic soil. (Zones 3–8)

S. menziesii (Alaskan burnet) grows 60 cm–1.2 m tall and 30–60 cm wide, bearing dark green to grey-green foliage and spikes of deep maroon flowers in summer. **'Dali Marble'** has attractive green foliage

S. menziesii

with white margins and bears red-purple flower spikes in late summer and fall. It is hardy to zone 3.

S. officinalis (greater burnet) has green foliage and small spikes of dark purple-red flowers that bloom all summer. Plants usually grow 60 cm–1.2 m tall and 60 cm wide. This species tolerates some drought as well as poor or high-alkaline soil. It self-seeds generously. **'Tanna'** is a compact selection, growing 30–46 cm tall and wide. (Zones 4–8)

Problems & Pests

Plants are generally problem free but can be afflicted by leaf spot.

S. officinalis

Buttercup

Ranunculus

Height: 5–90 cm • **Spread:** 20–90 cm • **Flower colour:** yellow, white •
Blooms: summer to fall • **Hardiness:** zones 3–8

Buttercups boast bright, shiny flowers and attractive foliage. They look great around a pond, where the soil tends to remain cooler. Some members of this genus can be invasive and weedy, including *R. repens*. Its cultivars are thought to be less aggressive but are still spreading plants.

Planting

Seeding: start fresh, ripe seeds in containers in a cold frame
Transplanting: spring
Spacing: 30–90 cm

Growing

The buttercups listed below grow well in **full sun, partial shade** or **light shade** in **evenly moist, well-drained** soil. Buttercups adapt to a range of soil types but prefer a cool rootzone. Water thoroughly before and during flowering, but allow the soil to dry out a bit once the blooming has finished. Divide in spring or fall.

Tips

Use the taller varieties in beds and borders, and as accents around water features. Use the ground-hugging creepers in rock and alpine gardens, in containers, along stone walls and for edging.

Recommended

R. acris (tall buttercup, yellow bachelor's button, meadow buttercup) is a narrow, clump-forming perennial, 60–90 cm tall and 20–30 cm wide, with wiry stems and deeply lobed leaves. Bright yellow flowers bloom in early to mid-summer. **Var. florepleno** is commonly available. It bears double, rosette-like, bright yellow blossoms.

R. ficaria (celandine buttercup, lesser celandine, pilewort) bears single, cup-shaped, bright yellow blossoms in spring. It grows only 5 cm tall and spreads 30–46 cm. The glossy green leaves are marked with hints of silver and gold. **'Albus'** produces single, creamy white flowers. **'Brazen Hussy'** bears bright yellow flowers and purple-black foliage, and **'Double Mud'** has pastel yellow flowers with brown markings resembling mud stains. (Zones 4–8)

R. repens (creeping buttercup) is a vigorous, somewhat invasive spreader with lobed foliage and bright yellow flowers. Plants grow 15–30 cm tall and 60–90 cm wide but can grow larger in ideal conditions. **'Buttered Popcorn'** has single yellow flowers and bright, chartreuse variegated, ornate foliage. (Zones 4–8)

Problems & Pests

Powdery mildew, slugs, aphids and spider mites can all become problematic.

The plant sap of buttercups may cause an allergic reaction in some people, but that same feature discourages animal browsing.

A. repens

Calamint

Calamintha

Height: 30–60 cm • **Spread:** 46–75 cm • **Flower colour:** pink, mauve, white •
Blooms: mid-summer to fall • **Hardiness:** zones 5–8

Calamint is a member of the mint family and has wonderful, minty foliage, especially when brushed against or crushed. Calamint self-seeds, especially where the plant is growing well, and you might find plants popping up in unwanted places in your landscape. Calamint is worth trying in the colder hardiness zones with good winter protection. The plants attract bees and butterflies.

Planting

Seeding: start seeds in containers in a cold frame or indoors in spring
Transplanting: after the last frost date
Spacing: 30–60 cm

Growing

Calamint grows well in **full sun to partial shade** in **moist, well-drained** soil. The plant is drought tolerant when established, and the soil can be allowed to dry before watering. Divide in early spring. Deadheading is required to mitigate self-seeding.

Tips

Calamint is best placed where you can enjoy the wonderful fragrance, such as along paths and at the edge of beds and borders. Use in woodland, cottage, meadow and rock gardens. Calamint can also be used as a groundcover.

Recommended

C. grandiflora is a bushy, mounded plant that grows 30–60 cm tall and 46–60 cm wide and spreads slowly by rhizomes. It has dark green, aromatic foliage and produces pink to light lavender blooms all summer long.

C. nepeta (lesser calamint, nepitella) is an upright, bushy plant with shiny, dark green foliage that grows 46–60 cm tall and 46–75 cm wide. It blooms from mid-summer to fall, bearing intricate, lavender to white flowers in abundance.

Problems & Pests

Powdery mildew may occur.

C. nepeta

The leaves make a refreshing tea. Pick the leaves as the plant is starting to bloom, then use sparingly fresh or dried. The fresh or dried leaves are also nice in a potpourri.

A. nepeta variety

Carolina Lupine

Bush Pea, False Lupine
Thermopsis

Height: 60 cm–1.5 m • **Spread:** 60 cm • **Flower colour:** dull to bright yellow •
Blooms: late spring to early summer • **Hardiness:** zones 3–8

Carolina lupine is very similar to lupine (*Lupinus*). Give it full sun in the cooler climates and partial shade in the hot, humid parts of the country. The foliage can become ratty after blooming and should be cut back, but once established, this is a tough plant, so don't let ratty foliage fool you into thinking it is not happy. You'll enjoy these pea flower blooms for one month every year.

Planting

Seeding: sow fresh seeds in a cold frame in fall or outdoors in spring; can also be started indoors in early spring
Transplanting: spring
Spacing: 46–60 cm

Growing

Carolina lupine grows well in **full sun to partial shade** in **fertile, well-drained, loamy** soil, but it adapts to a variety of soil conditions. This plant is drought and heat tolerant when established. Staking may be required in windy sites. An organic mulch reduces rootzone moisture loss and helps keep the roots cool.

Carolina lupine dislikes being divided, moved or having its roots disturbed. Propagation is by seeds or by carefully moving the young seedlings that grow at the base of the mother plant. Break open dried seedpods in fall to collect the fresh seeds.

Tips

Carolina lupine looks great when planted en masse. It is at home in a wildflower garden and also works well in the middle or at the back of perennial or mixed beds and borders. The flowers are excellent for cutting.

T. rhombifolia

Recommended

T. villosa (*T. caroliniana*) is a clump-forming, long-lived, native perennial that spreads slowly by rhizomes. It has thick, minimally branched stems and attractive, green, pea-like foliage. It bears lupine-like clusters of downy, dull to bright yellow pea flowers from late spring to early summer.

Problems & Pests

Carolina lupine is mostly pest free but may experience powdery mildew, leaf spot, aphids and slugs.

T. caroliniana

Cortusa

Alpine Bells
Cortusa

Height: 15–30 cm • **Spread:** 15–30 cm • **Flower colour:** pink, violet, white
Blooms: late spring to early summer • **Hardiness:** zones 5–8

This wonderful, little woodland plant is a member of the primrose family and enjoys the same habitats. The interesting foliage provides some coarse texture to the garden, and the delicate, nodding bells brighten up a woodland floor. Cortusa self-seeds where it is happy and is definitely worth trying in zone 4.

Planting

Seeding: sow seeds in situ or in a cold frame as soon as they are ripe; the seeds germinate best with exposure to light
Transplanting: spring
Spacing: 15–23 cm

Growing

Cortusa grows well in **partial to light shade** in **average to fertile, humus-rich, evenly moist, well-drained** soil. It will not thrive in hot and dry locations but may tolerate short dry periods. Make sure to protect the crowns with a thick layer of mulch for winter. When necessary, divide in spring.

Tips

Cortusa is suitable for woodland settings, shady borders, alpine gardens and shaded rock gardens.

Recommended

C. matthioli forms small clumps of attractive, rounded to kidney-shaped, lobed, coarsely toothed, hairy, medium to dark green foliage. In late spring to early summer, it bears small clusters of nodding, deep pink, fuchsia or white bell-shaped flowers held above the foliage on reddish stems.

Problems & Pests

Slugs and snails enjoy cortusa.

Cortusa is native to mountainous woodlands from Europe to Asia.

C. matthioli (both photos)

Crocosmia

Crocosmia

Height: 46 cm–1.2 m • **Spread:** 30–46 cm • **Flower colour:** red, orange, yellow • **Blooms:** mid- to late summer • **Hardiness:** zones 5–8

I would love a stand of crocosmia, but all the suitable places in my landscape are already occupied, and I live in zone 3. Gardeners in the colder areas of our country might have success if they can find the right microclimate. The best location is a protected, sunny corner of the garden. Savvy gardeners in the suitable climates are blessed to enjoy the gorgeous flowers and attractive foliage. Hummingbirds will love you for planting the fiery 'Lucifer.'

Planting

Seeding: start indoors or outdoors in early spring
Transplanting: spring
Spacing: 20 cm

Growing

Crocosmia prefers **full sun**. The soil should be of **average fertility, humus rich, moist** and **well drained**. Plant in a protected area and provide a good mulch of shredded leaves or other organic matter in fall to protect the roots from fluctuating winter temperatures. Divide in spring before growth starts, every two to three years, when the clump is becoming dense. Overgrown clumps produce fewer flowers.

Tips

This attractive, unusual plant creates a striking display when planted in groups in a herbaceous or mixed border. It also looks good planted next to a pond, where the brightly coloured flowers can be reflected in the water.

Recommended

C. x *crocosmiflora* is a spreading plant with long, strap-like leaves. It grows 46–90 cm tall, and the clump spreads about 30 cm. One-sided spikes of red, orange or yellow flowers are borne in mid- and late summer. **'Citronella'** ('Golden Fleece') bears bright yellow flowers.

C. 'Lucifer' is the hardiest of the bunch and bears bright scarlet red flowers. It grows 90 cm–1.2 m tall, with a spread of about 46 cm.

C. 'Lucifer'

Problems & Pests

Occasional trouble with spider mites can occur. Hose the mites off as soon as they appear. Browsing deer may also be a problem.

C. x *crocosmiflora* cultivar

Culver's Root

Bowman's Root
Veronicastrum

Height: 1.2–1.8 m • **Spread:** 46–90 cm • **Flower colour:** white, pink, purple-blue •
Blooms: mid-summer to fall • **Hardiness:** zones 3–8

Culver's root adds a distinctive, lofty, vertical accent to a composition of
perennials or prairie plants. These stately plants may reach as high as 2 m,
but regardless of how tall they grow, they are always attractive. *Veronicastrum*
is native to meadows and moist thickets from Manitoba to the east coast and
is easy to grow in adequately damp soil. It also supplies good cut flowers for
arrangements.

Planting
Seeding: start seeds in a cold frame in fall
Transplanting: spring
Spacing: 46–60 cm

Growing
Culver's root grows well in **full sun** or **partial shade**. The soil should be **average to fertile, humus rich** and **moist**. Divide plants in spring or fall when clumps appear to be thinning in the middle or when plants appear to be growing less vigorously.

Tips
Use culver's root at the back and middle of a mixed or herbaceous border. It makes a lovely addition to open woodland gardens, meadow plantings, cottage-style gardens and in the moist soil near a water feature. Its late-season blooming in pastel shades is a welcome change from the yellows, oranges and golds that tend to dominate at this time of year.

V. virginicum forma *album*

Recommended
V. virginicum is an upright perennial with whorls of deep green leaves on branching stems. Plants grow 1.2–1.8 m tall with a spread of 46–90 cm. Spikes of fuzzy-looking white, pink or purple-blue flowers are produced from mid-summer to early fall. **'Apollo'** bears light lilac flowers. ***V. v.* forma *album*** ('Album') bears white flowers. ***V. v.* forma *roseum*** bears pink flowers. ***V. v.* forma *roseum* 'Pink Glow'** bears pale pink flowers. **Var. *sibiricum*** is a super-hardy variety with long, lavender-blue flower spikes.

Problems & Pests
Problems with powdery or downy mildew and leaf spot can occur.

V. var. *sibiricum*

Cyclamen

Cyclamen

Height: 5–20 cm • **Spread:** indefinite • **Flower colour:** white, pink, red • **Blooms:** late winter to early spring • **Hardiness:** zones 5–8

Cyclamen's vivid flowers bloom at an unlikely time. The flower buds arise in fall to late winter, opening to reveal the beautiful pink, white and magenta flowers. The attractively marked foliage, ranging from dark green with varied patterns of marbled silver to a frosted pewter, arrives later. Many hybrids have fragrant flowers. Planting them under trees allows the tree roots to absorb excess moisture.

Planting

Seeding: soak fresh seeds in water for at least 12 hours; rinse before planting in a cold frame in fall; keep seeds in the dark until they germinate
Transplanting: plant tubers during cyclamen's dormant season or, with the fall or winter flowering varieties, after they have finished flowering
Spacing: 15 cm

Growing

These tuberous perennials grow best in **light** or **partial shade**. The soil should be of **average fertility, humus rich** and **well drained**. Good winter drainage is essential to prevent the tubers from rotting.

Tips

Cyclamen makes a lovely addition to a woodland or shaded garden. It can also be planted under shrubs and in shaded rock gardens.

Planting depth is not critical for these tubers because, like daffodils, they develop contractile roots and pull themselves down to the proper depth, depending on your garden's conditions; planting them 5–8 cm deep is a good rule of thumb.

Recommended

C. coum forms a small clump of rounded leaves. It grows 5–20 cm tall and slowly spreads by seed to form a colony of variable size. Flowers in shades of pink and magenta to white are borne from late fall through late winter, later than *C. hederifolium.*

C. hederifolium (ivy-leaved cyclamen) forms a low clump of triangular to

C. coum

heart-shaped leaves veined or marbled with silver, grey or bronze markings. It grows 10–15 cm tall and spreads 15–20 cm. Flowers in shades of pink or sometimes white are produced from fall to early winter and emerge before the new leaves. Plants usually go dormant in summer.

Problems & Pests

Spider and cyclamen mites as well as vine weevils can be a problem. Mice and squirrels may eat tubers or dig them up.

C. hederifolium

Daffodil

Narcissus

Height: 10–60 cm • **Spread:** 10–30 cm • **Flower colour:** white, yellow, peach, orange, pink; often bicoloured • **Blooms:** early to late spring • **Hardiness:** zones 3–8

Plant daffodils where the fading, yellowing or browning foliage won't mar your garden, such as in a meadow, at the base of large deciduous shrubs or in a thick perennial border. Shrubs that leaf out later in spring shout for a setting of colourful blossoms. Once the daffodil blossoms fade, the shrubs leaf out and hide any unsightly daffodil foliage.

Planting

Seeding: sow freshly ripened seeds into a cold frame in fall; plants can take five to seven years to begin flowering

Transplanting: plant bulbs in fall
Spacing: 10–20 cm

Growing

Daffodils grow best in **full sun** or **light, dappled shade**. The soil should be **average to fertile** and **well drained**. Keep the soil moist during growth and flowering, but allow it to dry out a little when the plants are dormant. Divide bulbs every four or five years, or when the flowers decrease in size and number in early summer and the foliage has thoroughly declined, or in early fall when you normally plant bulbs. Make sure to mark the bulbs' location so you can find them if the leaves disappear.

Plant bulbs when the weather cools in fall. As a guideline for how deeply to plant the bulb, measure it from top to bottom and multiply that number by three. Select the largest bulbs for the type you are growing. Ensure they are blemish-free with no soft spots.

Tips

Plant daffodils where they can be left to naturalize—in the light shade beneath a tree, in a woodland garden and in mixed beds and borders. Left undisturbed, they will multiply, making a more beautiful display each spring.

To keep your daffodil show looking fresh and to prevent seed production from sapping the bulb's energy, deadhead before the ovary (just below the petals) swells. Remove the entire stem but not the leaves until they turn yellow and begin to wither. Do not braid or tie them up with rubber bands.

Recommended

Many species, hybrids and cultivars are available. Flowers can be 4–15 cm across and may be solitary or borne in clusters. There are currently 13 flower-form categories. Choose dwarf varieties for small gardens.

N. **'Accent'** grows about 46 cm tall and bears flowers with crisp white perianths and large, sunproof, upfacing, salmon pink coronas. This plant is a vigorous performer.

N. 'Bravoure'

N. 'Ice Follies'

N. 'Actaea' grows 41–46 cm tall and has fragrant flowers that have snow white perianths with broadly over-lapping petals and small flat yellow coronas edged with bright red. It is the latest-flowering of all narcissus and one of the best for naturalizing. Plant it with Spanish bluebells for a classic combination in a shady spot in your May garden.

N. 'Bravoure' grows 36–46 cm tall and bears flowers with an unusual, long "stovepipe" of a corona and white, overlapping petals on its 13 cm wide perianth. This plant is an award winner and is very elegant.

N. 'Lemon Glow'

N. 'Decoy' grows 30–46 cm high and bears flowers with white perianths and rich, dark coral-pink coronas—a stunning and unusual colour.

N. 'Ice Follies' grows 41–46 cm tall and bears flowers whose white perianth backs a large, funnel-shaped, frilled yellow corona that fades to white as it matures. It is one of the easiest and most successful bulbs for naturalizing.

N. 'Intrigue' grows 25–30 cm tall and is a reverse bicolour daffodil, having deep yellow perianths and pale creamy white coronas. This plant is quite prominent in the garden.

N. 'Jenny' is a perennial award winner that grows 25–30 cm tall and bears flowers with gracefully recurved white perianths and slender, primrose-yellow coronas that ripen to creamy white.

N. 'Tete-a-Tete'

N. 'Kaydee' grows 20–30 cm tall and bears flowers with pure white, swept-back perianths and vivid salmon-pink coronas.

The cup in the centre of a daffodil flower is called the corona, and the group of petals that surrounds the corona is called the perianth.

N. 'Lemon Glow' grows 20–56 cm high and has won awards for its unusual pale primrose colour that fades to milky white. Its corona has a ruffled, darker yellow rim. Huge and robust, this plant always engenders admiration from garden visitors.

N. 'Mondragon' grows 30–46 cm tall and bears flowers with golden-yellow perianths and ruffled tangerine-coloured coronas, and it has a subtle apple scent.

N. *poeticus* 'Plenus' grows 41–46 cm tall and bears richly fragrant, snowy white double flowers. It is one of the few daffodils that shows up on almost every daffodil list from 1601 through to the catalogues of the early 1900s and can still be found today.

N. 'Stainless' grows 46–60 cm tall and bears flowers with ivory white perianths and luminous white, green-eyed coronas.

N. 'Stratosphere' is a vibrant, late-blooming jonquil hybrid that grows 66 cm tall, so it can be planted farther back in the border. It blooms in clusters of three, with yellow perianths and orangey yellow coronas. This plant is remarkably long lasting in the garden as well as when cut and brought indoors.

N. 'Tete-a-Tete' is a popular, fragrant miniature, about 15 cm tall, whose bright yellow flowers in clusters of 1–3 blooms are early and prolific. Great for naturalizing, they are also easy to force (getting them to bloom

N. 'Stainless'

before they would normally). In fact, the thousands I grow were purchased in bloom at a supermarket, six per pot, and planted out after they faded. Nine or so pots yielded scores of flowers within a few years.

Problems & Pests

Possible problems include basal rot, bulb rot, nematodes, bulb scale mites, slugs and large narcissus bulb flies. Also, flower blast causes buds to turn brown and dry up before opening. The cause is unknown but is thought to be weather related. Bulbs with this symptom year after year can be dug up and discarded.

Daffodils have triangular stems that help them turn their backs on the breeze to avoid breaking off while remaining upright. This makes them dance about more than other flowers, including tulips.

Like cyclamen bulbs, daffodil bulbs have contractile roots and will eventually adjust themselves to the proper soil depth.

Edelweiss

Leontopodium

Height: 15–30 cm • **Spread:** 25 cm • **Flower colour:** yellow flowers, silvery white bracts • **Blooms:** summer • **Hardiness:** zones 3–6

It is difficult to not think of the song "Edelweiss" from the movie *The Sound of Music* when dealing with this plant. The flowers smell like honey, and the whole plant is wonderful to touch. It makes you want to sing! The leaves and stems are covered with white hairs that protect the plant from cold and intense ultraviolet sunlight, which is useful to these tough little plants. They grow in rougher areas above treeline and are native to European and Western Asian mountain ranges.

Planting

Seeding: sow ripe seeds in a cold frame or directly into beds after the last frost
Transplanting: spring, summer
Spacing: 25 cm

Growing

Edelweiss thrives in **full sun,** but it tolerates partial sun. The soil should be **neutral to alkaline** and very **well drained**, which can be accomplished by cultivating in a soil-based potting mix or good compost. Divide in spring when necessary. Freshly divided plants may be slow to return to vigour. Edelweiss is extremely drought tolerant.

Tips

Edelweiss stands out in mixed beds and borders and remains small enough to be planted in alpine or rock gardens. The flowers can be used in dried floral arrangements.

Recommended

L. alpinum produces small mounds of long, silvery grey leaves. Grey,

L. alpinum

softly hairy stems are tipped with tiny, yellowish white flowers atop fuzzy, silvery white, pointed bracts.

Problems & Pests

Slugs and snails may cause the foliage to become unsightly.

The plant's genus name comes from the Greek words leon, *meaning "lion," and* podion, *meaning "paw," in reference to the shape of the flowers.*

L. album

Fairy Bells

Disporum

Height: 15–60 cm • **Spread:** 60 cm or more • **Flower colour:** white, pale yellow • **Blooms:** late spring to early summer • **Hardiness:** zones 4–8

Fairy bells is a delightful plant, but be aware that it can spread, especially where it is happy. Use this trait to your advantage and let it ramble to form colonies in your woodlands and wild areas. Fairy bells is somewhat rare, and you might have to do some searching to find it.

Planting

Seeding: start fresh, ripe seeds in a cold frame in fall
Transplanting: spring, after the risk of frost has passed
Spacing: 60–90 cm

Growing

Fairy bells prefers **partial to light shade** in **moist, humus-rich, well-drained** soil. Divide rhizomes in early spring while plants are dormant.

Tips

Fairy bells makes a lovely addition to shaded beds and borders, and especially when left to naturalize in woodland gardens.

D. sessile (both photos)

Recommended

D. sessile has upright to arching, sparse branched stems and mid- to dark green foliage with wavy margins. It bears small clusters of hanging, tube-like, white to very pale yellow flowers with green tips, from late spring to early summer. The purple to black berries provide fall interest. **'Variegatum'** has green foliage that is striped and margined with white.

Problems & Pests

Vine weevils, slugs and leaf spot are potential problems.

False Indigo

Baptisia

Height: 60 cm–1.5 m • **Spread:** 60 cm–1.2 m • **Flower colour:** white, purple-blue • **Blooms:** late spring, early summer • **Hardiness:** zones 3–8

False indigo is an elegant, tough-as-nails perennial that should be grown far more often than it is. Clumps of *B. australis* can grow as big as some shrubs! A member of the pea family, this plant sends up slender stems of indigo-blue flowers in late May to June. The long, oval foliage of *B. australis* has a bluish tint; foliage of all varieties remains attractive all season. This plant can be left undisturbed for many years.

Planting

Seeding: sow seeds indoors in early spring, or direct sow in late summer; protect plants for the first winter
Transplanting: spring
Spacing: 60–90 cm

Growing

False indigo prefers **full sun** but tolerates partial shade. Too much shade results in lank growth that causes the plant to split and fall. The soil should be of **average** or **poor fertility, sandy** and **well drained**. *B. alba* and *B. bracteata* are more tolerant of dry soils than *B. australis*, which occurs on stream banks. Divide carefully in spring, only when you want more plants.

B. alba

B. australis

B. australis (all photos)

B. australis 'Purple Smoke'

Staking may be required if the plant is not getting enough sun. Rather than worry about staking or having to move the plant, place it in the sun and give it lots of space to spread.

The hard seed coats may need to be penetrated before the seeds can germinate. Scratch the seeds between two pieces of sandpaper before planting them.

Tips

False indigo can be used in an informal border or cottage-type garden. It is attractive when used in a natural planting, on a slope or in any well-drained, sunny spot in the garden.

Recommended

B. alba is an erect, bushy plant growing 60 cm–1.2 m tall and 60 cm wide.

In early summer, it produces spike-like, open clusters of creamy white flowers that sometimes have purple-tinged upper petals. (Zones 4–9)

B. australis (plains false indigo) is an upright or somewhat spreading, clump-forming plant, 90 cm–1.5 m tall and 60 cm–1.2 m wide, that bears spikes of purple-blue flowers in early summer.

B. bracteata is an erect to spreading, softly hairy plant with arching stems, growing 60–62 cm tall and wide. It produces erect to semi-pendent clusters of white to creamy white flowers in late spring to early summer.

Problems & Pests
Minor problems with mildew, leaf spot and rust can occur.

Foamy Bells

Heucherella
x *Heucherella*

Height: 20–46 cm • **Spread:** 30–46 cm • **Flower colour:** white, pink; grown for foliage • **Blooms:** spring to mid-summer • **Hardiness:** zones 3–8

Foamy bells are most often grown for their foliage. The bold forms blend beautifully into shady beds and borders, cottage gardens or in a wild, woodland setting. The large, colourful leaves contrast well with other variegated plants and plants with brightly coloured flowers. These attractive plants show great promise with continued hybridizing. All foamy bells are sterile. Two popular selections are 'Stoplight' and 'Sunspot,' both with deep red hearts on yellow leaves, which are great for brightening up a shady path.

Planting

Seeding: not recommended
Transplanting: spring or fall
Spacing: 30–46 cm

Growing

Foamy bells grow well in **full sun**
and **partial** or **light shade** but tolerate full shade. They seem to prefer
morning sun and afternoon shade.
The soil should be **fertile, humus
rich, neutral to acidic, moist** and
well drained. They don't perform as
well when they have to compete with
other heavy feeders or when their
soil is allowed to dry out for pro-
longed periods during hot weather.
Divide in spring or fall.

Tips

Foamy bells make a lovely addition
to moist borders and open woodland
gardens. The brightly coloured leaves
and low growth habit make them a
much sought after groundcover
plant, and the foliage contrasts
attractively with a wide variety of
other perennials.

Recommended

x *H.* 'Bridget Bloom' forms into a
mound of dense, lobed foliage lightly
mottled with lighter shades of green.
Dark, wiry stems support pink
flower spikes. Plants grow 30–46 cm
tall and 30 cm wide.

x *H.* 'Burnished Bronze' forms a
mound of bronzy purple leaves.
Dark purple stems contrast with the
light pink flowers. It grows 20–46 cm
tall and spreads about 46 cm.

x *H.* DAYGLOW PINK has brilliant pink
flowers and deeply lobed, light to
mid-green leaves with chocolate-
coloured veins. Plants grow 30–46 cm
tall and wide.

x *H.* 'Kimono' has deeply lobed, silvery green leaves with mottled purple veining. It bears pale pink flowers
and grows 20–46 cm tall with an
equal spread.

x *H.* 'Stoplight' has chartreuse leaves
with red centres. Flowers are white.
Plants grow 30–46 cm tall with an
equal spread.

x *H.* 'Sunspot' has yellow-green foliage with red central veining. Flowers
are pink. Plants grow 20–46 cm tall
and spread about 46 cm.

Problems & Pests

Foamy bells rarely suffer from any
problems.

x *H.* 'Bridget Bloom'

Fringe Cups

Tellima

Height: 60–90 cm • **Spread:** 30–60 cm • **Flower colour:** greenish white or chartreuse, fading to pink • **Blooms:** mid-spring to early summer • **Hardiness:** zones 4–8

Fringe cups spread by self-seeding and by thick rhizomes and can spread somewhat quickly if they are in ideal conditions. They are great plants for the floor of your woodland or in a wild meadow, where their spreading habit can be encouraged. Keeping mature plants a little on the dry side helps mitigate the rate of spread. These easy-care plants are native to British Columbia.

Planting

Seeding: start ripe seeds in containers in a cold frame in spring or fall
Transplanting: spring, after the risk of frost has passed
Spacing: 46–60 cm

Growing

Fringe cups grow best in **partial to light shade.** The plants tolerate full shade and might be okay in full sun in the colder summer areas, provided the soil remains moist. Ideally, the soil should be **moist, humus rich** and **slightly acidic,** but fringe cups tolerate a range of soils. Mature fringe cups are also tolerant of drought and occasional mild flooding. Divide in early spring or early fall. Deadhead the entire flowering stems to keep the plants tidy and to reduce self-seeding.

Tips

Fringe cups work well as accents or when planted en masse in beds, borders, woodland gardens and rock gardens. They can also be used as a groundcover.

Recommended

T. grandiflora forms low rosettes of attractive, heart-shaped to rounded, hairy, rich evergreen foliage with scalloped margins. The flowering stems rise well above the foliage, and each stem bears a plethora of fragrant, greenish white, fringed, cup-shaped flowers that fade to pink with age. **'Forest Frost'** has green foliage with frosty silver variegation. The green part of the foliage turns burgundy in fall. It bears chartreuse flowers.

T. grandiflora

Problems & Pests

Slugs are potential problems, and fringe cups may also get powdery mildew.

The flowers bloom from the bottom to the top of the flowering stems, and you get both greenish white flowers and pink flowers on the same stem.

T. grandiflora is great for attracting butterflies and birds, including hummingbirds. Deer, however, do not enjoy this plant.

Gas Plant

Burning Bush, Dittany, Fraxinella
Dictamnus

Height: 46–90 cm • **Spread:** 30–90 cm • **Flower colour:** white, pink, pink-purple with darker veins • **Blooms:** early summer • **Hardiness:** zones 3–8

Gas plants are easy-to-grow, long-lived, trouble-free perennials that require little care. They are prized for their dark green, glossy foliage as well as their flowers, which have attractive veining. Gas plant is very slow growing and can take two to three years to become established.

Planting

Seeding: sow seeds in the garden or in containers and set them outside in winter in colder climates; the seeds need cold treatment for germination
Transplanting: spring
Spacing: 30–60 cm

Growing

Gas plant prefers **full sun** but tolerates partial shade. It prefers areas with cool nights. The soil should be **average to fertile, dry** and **well drained**.

This plant takes several years to become established and should not be divided because it resents being disturbed. It may not flower until it has become well established.

Tips

Gas plant makes a good addition to a border. You won't have to do much maintenance once it is established, so feel free to plant it in an area that is hard to reach.

Although the foliage has an appealing scent, avoid planting gas plant where you will brush against it. The oils in the foliage can cause photodermatitis, which means that the oils themselves may not cause irritation, but in combination with exposure to sunlight, they can cause rashes, itching and burning. The problem is often worse for people with fair skin.

Recommended

D. albus *(D. fraxinella)* is a large, clump-forming plant with lemon-scented leaves. The plant is long

D. albus var. *purpureus* (both photos)

lived, but it takes a few years to become established. It bears spikes of pink-veined, white or pink flowers. **Var.** *purpureus* ('Purpureus') has light pink-purple flowers with darker purple veins.

Germander

Teucrium

Height: 15–60 cm • **Spread:** 15–46 cm • **Flower colour:** pink to purple • **Blooms:** late spring to early summer • **Hardiness:** zones 4–8

Germander is an excellent, but underused, perennial. Actually it's a subshrub, which is a woody-based perennial that can regenerate from the roots if the top dies back over winter. It is worth planting in zones 4 and 5, even if it takes a little extra maintenance to provide winter protection. Germander is often used as a boxwood substitute and can be sheared into the type of low, tight hedge you might see in a formal garden or knot garden. The plants are drought tolerant when established and are good for attracting bees and butterflies into your garden.

Planting

Seeding: sow fresh ripe seeds in containers in a cold frame
Transplanting: spring or fall
Spacing: 15–46 cm

Growing

Germander grows best in **full sun** in **well-drained, moderately moist to dry, neutral to alkaline** soil of **average fertility**. Germander tolerates partial shade but may get leggy. It tolerates poor soils. Provide a sheltered location and winter protection in zones 4 and 5. To propagate plants, divide them when they are dormant in early spring or fall, or take cuttings in early to mid-summer.

Pinch or cut back the plants after flowering to promote bushy, compact growth, or if they are leggy or scruffy looking. They can be cut to 5 cm above the ground in early spring. Germander can also be lightly sheared regularly throughout the growing season when used as a short hedge.

Germander can spread aggressively, so you need to be attentive about keeping it in bounds. Plant it in deep, bottomless containers or use deeply buried barriers to stop the spread. The spreading trait is useful for erosion control on slopes.

Tips

Germander can be used for edging, low hedges and as a groundcover. It looks good alone or when planted in containers, knot and formal gardens, rock gardens and herb gardens.

Recommended

T. chamaedrys (wall germander) is an upright, evergreen or deciduous subshrub that grows 30–60 cm tall and 30–46 cm wide. It has aromatic, oval-shaped, shiny, dark green foliage. Spikes of tubular, two-lipped, pink to purple flowers bloom in late spring to early summer, but bloom time varies with climate. 'Nanum' ('Prostratum') forms compact mounds 15–30 cm tall and wide. 'Summer Sunshine' has bright yellow leaves in spring that turn yellow-green in summer.

Problems & Pests

Problems are infrequent, but rust, mildew, mites and leaf spot may afflict germader. It may suffer serious injury or death in severe winters. The plant is resistant to deer and honey fungus.

T. chamaedrys

Globeflower

Trollius

Height: 30–90 cm • **Spread:** 40–60 cm • **Flower colour:** yellow, orange •
Blooms: spring to early summer • **Hardiness:** zones 3–7

Do you have clay soil that remains moist all summer? Globeflower is an
unusual plant that thrives in such conditions. This moisture-loving plant is
at home in formal and informal plantings and is superb as a pool or streamside specimen plant. Clay soils should be amended with copious amounts of
high-quality compost, but if that task is too troublesome, just use the compost as mulch around the plants.

Planting

Seeding: sow fresh, ripe seeds in a cold frame in fall or spring; seeds may take up to two years to germinate
Transplanting: spring or fall
Spacing: 46–60 cm

Growing

Globeflower prefers **partial shade** but tolerates full sun if enough moisture is provided. Plants prefer cool, moist conditions and do not tolerate drought. The soil should be **fertile** and **heavy** and should not be allowed to dry out. Globeflower can be planted in well-drained soil as long as the soil remains moist. Prune out any yellowing leaves in summer. Division is rarely required but can be done in early spring or late fall.

Tips

Globeflower is the perfect plant for the side of a pond or stream. It naturalizes well in a damp meadow garden or bog garden and can be used in the border as long as the soil remains moist. Globeflower is long lasting as a cut flower.

Recommended

T. chinensis (Chinese globeflower) grows 60–90 cm tall and 46 cm wide. It produces yellow or orange bowl-shaped flowers with pronounced stamens. **'Golden Queen'** grows to 60 cm tall and has orange flowers.

T. x *cultorum* (hybrid globeflower) forms perfectly globe-shaped flowers. It grows 60–90 cm and 41–46 cm wide. **'Earliest of All'** grows 46–51 cm tall and produces pale yellow-orange

T. chinensis 'Golden Queen'

flowers earlier than the other globeflowers. **'Orange Princess'** has large, deep orange flowers on plants 30–60 cm in height.

T. europaeus (common globeflower) has dark green foliage and grows 60 cm tall and wide, bearing lemon yellow flowers that resemble those of buttercups. **'Superbus'** produces a plethora of 2.5–5 cm wide blooms.

Problems & Pests

Powdery mildew may cause occasional problems.

T. x *cultorum* 'Orange Princess'

Golden Hakone Grass

Japanese Forest Grass
Hakonechloa

Height: 25–46 cm • **Spread:** 60–90 cm • **Flower colour:** grown for its decorative, strap-like foliage; fall colour • **Blooms:** grown for foliage • **Hardiness:** zones 5–8

Not many grasses perform well in shade, but this one does. Try a combination of hellebore, hosta and ferns for a spectacular shady border. The denser the shade, the more lime green the foliage will turn, making a showier display. Colder zone gardeners are out of luck with this plant unless they have a spot with a really good winter microclimate.

Planting

Seeding: not recommended
Transplanting: spring
Spacing: 60–90 cm

Growing

Golden hakone grass grows well in **partial to light shade** in **moist, well-drained, moderately fertile, humus-rich** soil. The leaf colour is enhanced in partial to light shade. Golden hakone can tolerate full shade, but its foliage turns lime green under such conditions. It also tolerates full sun as long as the soil remains moist. Divide in spring. Avoid locations with cold, drying winds.

H. macra 'Aureola' (both photos)

Tips

Golden hakone grass works well at the edge of a woodland garden but is also effective in containers, beds, borders and rock gardens. It makes an excellent addition to larger gardens where it can be planted in drifts. The striking foliage cascades nicely over the edges of containers or low borders.

Recommended

H. macra is a slow-spreading, mound-forming perennial that has bright green, arching, grass-like foliage that turns deep pink in fall, then bronze as winter sets in. **'Alboaurea'** has bright yellow foliage with thin, green, vertical stripes. **'Aureola'** also has bright yellow foliage with narrow, green streaks; the foliage turns pink in fall. Yellow-leaved cultivars may scorch in full sun and lose their yellow colour in too much shade.

Problems & Pests

Golden hakone grass rarely suffers any problems.

Green and Gold

Golden Star
Chrysogonum

Height: 20–25 cm • **Spread:** 46–60 cm • **Flower colour:** bright yellow • **Blooms:** spring, then sporadically through summer • **Hardiness:** zones 5–8

Green and gold is a low-growing, shade-loving groundcover with bright yellow, foliage-hugging flowers that can be seen from metres away. The showy blossoms really brighten up a shady bed with their intensity of colour. Let the plants form colonies where they're happy. Green and gold is definitely worth trying in zone 4.

Planting

Seeding: start freshly ripened seeds in a cold frame in fall
Transplanting: spring
Spacing: 46–60 cm

Growing

Green and gold prefers **partial to full shade**. It the colder zones, it tolerates full sun and flowers less in full shade. This plant adapts to most **moist, well-drained** soils and benefits from the addition of **good compost** to the soil. Cut back ratty foliage or the entire plant in fall. Propagate plants by dividing in spring or fall. The daughter plants that grow from the nodes of the runners are easily transplanted.

C. virginianum (both photos)

Tips

This spreading perennial is often used as a flowering groundcover. It can be used for a blast of bright colour in woodland gardens and at the front of perennial and shrub borders.

Recommended

C. virginianum forms an attractive mat of toothed, coarse-textured foliage that spreads by runners that root at the nodes. Starry-shaped, bright yellow flowers bloom prolifically in early to mid-spring, with sporadic blooms throughout summer. A number of cultivars are available with dark green leaves, wider spreads, longer blooming periods or more vigorous growth habits.

Problems & Pests

Green and gold rarely suffers any problems but can be affected by powdery mildew.

Hair Grass

Deschampsia

Height: foliage 20–60 cm; flowers 60–100 cm • **Spread:** 30–60 cm • **Flower colour:** silvery purple, light bronze purple, bright yellow to golden yellow, pinkish to purple-tinged • **Blooms:** late spring to late summer • **Hardiness:** zones 3–8

Hair grass is another grass that does well in partly shady locations. The plant grows reasonably well in deeper shade, but the flowering will be noticeably reduced. It requires little maintenance when established, but be aware that it may self-seed if it likes where it is. The blades are quite narrow, and the whole plant has a fine texture, which can soften a planting of coarse-textured materials.

Planting

Seeding: sow seeds directly or in containers in a cold frame in spring or fall; do not cover the seeds
Transplanting: spring or fall
Spacing: 30–60 cm

Growing

Hair grass prefers **full sun to light shade** in **moist, humus-rich, acidic** soil but tolerates most soil types, including heavy clay. *D. caespitosa* likes heavier, moist soil. *D. flexuosa* likes well-drained, average to dry soil and is drought tolerant when established. Too much moisture in winter can be harmful. This plant is tolerant of full shade as long as it is not too deep.

Cut back flower stems in early spring before new growth begins. Divide in spring or fall.

D. flexuosa cultivar

D. caespitosa 'Goldstaub'

D. caespitosa 'Bronzeschleier'

D. caespitosa

Tips

Use hair grass in meadow, rock and woodland gardens, in borders and beds and as a groundcover. Keep the plant spacing tight when using it as a groundcover. Hair grass looks great as a specimen or when mass planted and is suitable for container planting. The flowering stems are attractive in fresh and dried arrangements. Tufted hair grass does well in bog gardens and beside ponds and streams. Wavy hair grass is good for a dry, shady location.

Recommended

D. caespitosa (tufted hair grass) is a dense, tuft-forming grass with arching, dark green foliage that grows 40–60 cm tall and 30–60 cm wide. Flowering stems often reach 90 cm tall and bear large, open clusters of silvery purple flowers in summer. **'Bronze Veil'** ('Bronzeschleier') is a slightly larger selection that reaches 1 m and has medium to dark green foliage and large plumes of light bronze purple flowers. **'Gold Veil'** ('Goldschleier') is another taller selection. It bears silver-tinged flowers that turn bright yellow. **'Gold Dew'** ('Goldtau') has dark green leaves and flowers that mature to golden yellow. It blooms later than other *D. cespitosa* selections. **'Northern Lights'** (variegated tufted hair grass) is grown for its upright, cream-and-green-striped foliage that is tinged pink in spring. Tufts grow 20–36 cm tall and 30–60 cm wide.

D. caespitosa cultivar

D. flexuosa (wavy hair grass, crinkled hair grass) forms loose, low, rounded tufts of slightly arching, medium to bright green foliage. Open, graceful clusters of silvery, pinkish or purple-tinged flowers are held above the foliage, blooming from late spring to early summer. The foliage grows 20–30 cm tall and 30–46 cm wide, and the flowering stems reach up to 60 cm in height. **'Tatra's Gold'** bears golden yellow foliage and bronze purple flowers in summer.

For 'Northern Lights' and other variegated plants, always remove all solid green foliage because it is usually more vigorous and can crowd out the variegated form.

Hair grass grows well at the edge of a woodland. The grass forms a symbiotic relationship with mycorrhizal fungi that live in the soils in and around mature forests.

Problems & Pests
Hair grass is usually free of pests and disease but may experience rust, caterpillars and sap-sucking bugs.

Hardy Orchids

Bletilla / Spiranthes

Height: 25–60 cm • **Spread:** 8–60 cm • **Flower colour:** pink, white •
Blooms: spring to early summer, fall • **Hardiness:** zones 5–8

All orchids are not the frail, tropical, greenhouse plants that need extra-special care and fussing. There are species of orchids everywhere, including in the arctic tundra. The only places one might not find orchids are in the middle of a desert or in aquatic habitats. A plethora of terrestrial orchids are native to North America, and *S. cernua* is native to Eastern Canada. *Bletilla* and *Spiranthes* are relatively easy to grow. Do not dig orchids from the wild; only purchase orchids from reputable merchants.

Planting
Seeding: not recommended
Transplanting: early spring, preferably while dormant
Spacing: 8–46 cm

Growing
Both species grow best in **partial shade** in a **sheltered location**. The soil should be **fertile, humus rich, moist** and **well drained**. Divide plants in early spring while they are dormant.

Tips
Both these plants make lovely additions to woodland gardens. *Spiranthes*, if grown in a boggy area, will spread to form an attractive colony. These orchids can also be used in shaded beds and borders as well as next to water features.

Recommended
B. striata forms a clump of long, deeply veined leaves. It grows 25–60 cm tall with an equal spread. Delicate sprays of magenta pink flowers are borne in spring and early summer. **Var. *japonica f. gebina*** bears creamy white flowers.

B. striata (both photos)

S. cernua (nodding ladies' tresses) grows up to 60 cm tall and spreads 8–15 cm, forming narrow, upright clumps of star-shaped leaves. It sends up spikes of white flowers with yellow centres in fall. *F. odorata* **'Chadd's Ford'** has larger fragrant flowers than the species.

Problems & Pests
Spider mites, aphids, whiteflies and mealybugs can cause trouble.

Hellebore

Christmas Rose, Lenten Rose
Helleborus

Height: 20–80 cm · **Spread:** 20–46 cm · **Flower colour:** white, green, pink, purple, yellow · **Blooms:** late winter, mid-spring · **Hardiness:** zones 4–8

Here's an unusual, low-growing woodland flower not widely grown in Canada but worth seeking out. The dark green, leathery foliage of *H. niger* appears in early spring followed by white, cupped, usually pendent flowers that flush slightly to pink. The key to success with hellebore is placement in a moist, cool, partly shaded location in soil that is neutral to slightly alkaline. Once *Helleborus* is happy with its surroundings, it will reward you with spring magic for many years to come. Hellebore is worth trying in the colder hardiness zones.

Planting

Seeding: not recommended; seeds are very slow to germinate
Transplanting: spring or late summer
Spacing: 30–46 cm

Growing

Hellebore prefers **light, dappled shade** in a sheltered site. It accepts a fair amount of direct sun if the soil is moist. The soil should be **fertile, moist, humus rich, neutral to alkaline** and **well drained**. Protect plants with mulch in winter, though in a mild winter you may find the flowers poking up through the snow in February. Divide in spring, after flowering or whenever plants become too crowded, or thin out in the centres. These plants self-seed, and the seedlings are variable. Freshen plants by removing spent leaves in late spring. Deadheading does not produce new blooms.

H. x hybridus cultivar (both photos)

Tips

Use these plants in a sheltered border or rock garden or naturalize in a woodland garden.

H. foetidus

H. niger

All parts of *Helleborus* species are **poisonous** and may cause intense discomfort if ingested. The sap may aggravate skin on contact. The leaf edges of some species are very sharp, so wear long sleeves and gloves when planting or dividing.

Recommended

H. foetidus (bearsfoot hellebore, stinking hellebore) is an upright, clump-forming plant. It grows 46–80 cm tall, spreads 30–46 cm wide and bears clusters of fragrant pale green flowers, often with purple margins. (Zones 5–8)

H. x *hybridus* (Lenten rose, Oriental hybrids) plants grow about 46 cm in height and spread. They are attractive and may be deciduous or evergreen. Plants bloom in late winter to spring in a wide range of flower colours, including white, purple, yellow, green and pink. Many plants sold as *H. orientalis* are hybrids. Trends include deeper-coloured flowers, picotees (with differently coloured petal margins), doubles and spotted flowers. (Zones 5–8)

H. niger (Christmas rose) is a clump-forming evergreen. It grows 30 cm tall and spreads 46 cm and bears white or pink-flushed flowers in early spring. This species is the most available and is easier to grow than *H. orientalis*.

H. odorus has leaves like most of the Oriental hybrids and luminescent, sweet to musky-scented, soft

H. foetidus

green 8 cm flowers. It is purport-
edly the best and the toughest of the
green-flowered hellebores. It grows
20–30 cm tall and wide.

Problems & Pests
Problems may be caused by aphids,
crown rot, leaf spot and black rot, and
by slugs when the leaves are young.

*Hellebores make superb cut
flowers. Place them high so you
can admire their pretty faces.*

*Unlike many plants that look
best planted en masse, hellebores
make exceptional specimen plants
to admire on their own.*

Holly Fern

Polystichum

Height: 30–60 cm • **Spread:** 30–90 cm • **Flower colour:** no flowers; grown for foliage • **Hardiness:** zones 3–8

What would a woodland garden be without ferns? *Polystichum*, commonly known as holly fern, is one of the 10 largest genera of ferns—there are about 260 species worldwide. A common species is the Christmas fern, an Eastern Canada native that can even be grown satisfactorily as a houseplant. It's undemanding, stays semi-green through winter and is highly adaptable to less than ideal conditions, unlike many of its ferny counterparts. The tassel fern is among the best of the evergreen ferns, retaining its fronds through the year and rarely requiring trimming.

Planting

Seeding: mature spores can be
started in a cold frame
Transplanting: spring or fall
Spacing: 30–60 cm

Growing

Both Christmas fern and tassel fern
grow well in **partial to full shade.**
The soil should be **fertile, humus
rich, moist** and **well drained.** In
spring before new fronds emerge,
trim off any that look worn out.
Divide rhizomes in spring to propagate your plants or carefully remove
offsets from the base of the plant.

P. acrostichoides (both photos)

Tips

These non-invasive ferns make
lovely additions to a moist woodland
garden or shaded water feature. They
can be mass planted in borders or
left to naturalize in largely unused,
shaded areas of the garden.

Recommended

P. acrostichoides (Christmas fern)
forms a circular clump of arching,
evergreen fronds. It grows 30–46 cm
tall and spreads 30–90 cm. Ideal for
a shady rockery, this fern tolerates
drier and sunnier places than most
ferns. This plant is native to eastern
North America.

P. polyblepharum (tassel fern, Japanese tassel fern) forms a circular
clump of dark green, glossy evergreen
fronds. Young fronds are covered in
golden hairs when they first begin to
unfurl and have a tassel-like appearance. The plants grow 30–60 cm tall
and spread about 60 cm. (Zones 5–8)

Problems & Pests

Holly ferns rarely suffer from any
problems.

*The use of its evergreen fronds to
decorate during the holidays gave
Christmas fern its common name.*

Hyssop

Hyssopus

Height: 30–60 cm • **Spread:** 60–90 cm • **Flower colour:** blue-purple, pink, white •
Blooms: mid-summer to early fall • **Hardiness:** zones 4–8

Hyssop is a wonderful, multi-use perennial. Its bushy, upright, fine-textured form combines well with coarser plants, and the long-lasting flowers will not clash with most of the coloured flowers you choose to plant them with. The fragrant foliage is used medicinally and as flavouring in food preparation, and the plant is said to repel certain insects. Hyssop self-seeds where it is happy. I know—they have dotted themselves through the full sun, southern exposure, minimum maintenance prairie bed at my home.

Planting

Seeding: start seeds in a cold frame in spring or fall
Transplanting: spring
Spacing: 60–90 cm

Growing

Hyssop grows well in **full sun** in **fertile, well-drained** soil. Begin with a starter plant rather than sowing from seed, especially in the colder zones, so it will reach a reasonable size in regions with shorter, cooler summers. Every few years, the rootball should be dug up in early spring to cut away and remove the old woody crown. Stem cuttings can be taken in summer before blooming occurs.

Tips

Use hyssop in perennial or mixed beds and borders and in the herb garden. It looks good when mass planted and as edging. Hyssop is a great container plant for window boxes and for locations on a hot deck or balcony. Hyssop is also attractive to bees, butterflies and hummingbirds.

Recommended

H. officinalis is a semi-evergreen shrub that acts like a perennial in most of Canada. It produces spikes of narrow, aromatic, mid-green leaves, topped with funnel-shaped, two-lipped, dark blue-purple flower spikes over a long period. **Forma albus** ('Albus') produces white flowers. **Subsp. aristatus** is a dense, compact selection. **Forma roseus** ('Roseus') bears pink flower spikes. Other good cultivars are available. Check with your local nursery or garden centre.

H. officinalis (both photos)

Problems & Pests

Insect pests or disease rarely bother hyssop.

The flowers and foliage are edible but pungent. An essential oil extracted from the plant is used to add fragrance to soap and perfume, and fresh or dried leaves and flowers can be added to a hot bath.

Ice Plant

Delosperma

Height: 2.5–15 cm • **Spread:** 20 cm to indefinite • **Flower colour:** pink-purple, pink, fuschia, magenta, yellow, orange-red • **Blooms:** late spring to fall • **Hardiness:** zones 4–8

Ice plant is a low-maintenance, low-growing, succulent perennial boasting very bright, eye-catching, aster-like flowers that open in bright sun. The plant is drought tolerant and absolutely needs good drainage. Any extra soil moisture may kill the plant, but with good drainage, this gem is quite hardy. Zone 4 might be pushing the limits of ice plant, but try it anyway.

Planting

Seeding: sow seeds in containers in a cold frame in late winter to early spring, or direct sow in fall; the seeds germinate easily in a damp paper towel
Transplanting: spring, after the risk of frost has passed
Spacing: 25–60 cm

Growing

Ice plant prefers to grow in **full sun; partial sun** is adequate but results in fewer flowers. The soil should be **well drained** and of **average fertility.** To prevent rot, avoid areas that are too wet. The plant will benefit from a site protected from excessive winter exposure that might dry out the foliage. A thick, dry, winter mulch is recommended. Ice plant propagates easily by stem cuttings in spring and summer or division in spring or fall.

Tips

Ice plant is frequently used in rock gardens, trough gardens, as edging in beds and borders and as ground-cover on rocky slopes for erosion control. It is a good candidate for use in containers with other xeric plants.

Recommended

D. cooperi is a creeping, succulent perennial with a mat-forming habit. The thick, stubby leaves are bluish green. Bright pinkish purple, aster-like flowers are produced in mid- to late summer. This species grows 2.5–8 cm tall and 46–60 cm wide. MESA VERDE bears pale coral to salmon-pink flowers. TABLE MOUNTAIN bears deep pink to fuchsia flowers.

D. floribundum 'Starburst'

D. floribundum is a clump-forming, semi-evergreen plant that grows 8–15 cm tall and 20–30 cm wide and has succulent, green to blue-green foliage. The bright pink-purple flowers have white centres and are held above the foliage. Plants begin blooming in late spring and continue until fall. **'Starburst'** has bright metallic pink to magenta blooms.

D. nubigenum is very similar in habit and appearance to *D. cooperi*, but the flowers are borne in fiery shades of yellow and orange-red and bloom in summer. It grows 5–10 cm tall and can spread indefinitely.

Problems & Pests

Plants may experience aphids and mealybugs.

D. cooperi TABLE MOUNTAIN

Indian Pink

Woodland Pinkroot
Spigelia

Height: 30–60 cm • **Spread:** 46 cm • **Flower colour:** red and yellow bicoloured • **Blooms:** spring and early summer • **Hardiness:** zones 5–8

Indian pink is an easy-to-grow perennial that does well just about anywhere in the garden. The bright, tubular flowers bloom over a long period and will catch the eye of your neighbours and passersby, as well as any local humming-birds. The dark green foliage is attractive throughout the growing season. In open woods, this plant can form large colonies but does not spread as aggressively in a home landscape.

Planting

Seeding: easily grown from seed; direct sow in late summer or fall
Transplanting: spring
Spacing: 30–46 cm

Growing

Indian pink prefers **light** or **partial shade** but tolerates full sun if the soil remains moist. The soil should be **fertile, moist** and **well drained**. Divide plants in spring.

Tips

Indian pink makes a nice addition to a sunny or shaded bed, border or woodland garden. It can be included in wildflower and native plant gardens. Try it with white bleeding hearts, ferns, columbines and mayapples for an attention-grabbing border.

S. marilandica (both photos)

Recommended

S. marilandica forms an upright clump. It grows about 60 cm tall and spreads about 46 cm. The tubular flowers are cherry red on the outside and vivid yellow inside.

Problems & Pests

Rare problems with leaf spot or powdery mildew can occur, otherwise this plant has no serious pests.

Inula

Inula

Height: 46 cm–2.4 m • **Spread:** 30 cm–1.5 m • **Flower colour:** yellow • **Blooms:** summer, fall • **Hardiness:** zones 3–8

Inulas are tough, low-maintenance, easy-to-grow plants that feature bright, long-lasting flowers with narrow, frilly petals and velvety foliage. Most species have large, bold foliage and look good surrounded by smaller, finer-textured plants. Inulas self-sow in ideal conditions, but not aggressively. The flowers attract butterflies and bees.

Planting

Seeding: start seeds in a cold frame in spring or fall
Transplanting: spring
Spacing: 30–90 cm

Growing

Grow inula in **full to partial sun** in **moist, well-drained** soil. The plants tolerate a variety of soils and can grow quite large in a fertile loam. They also tolerate winds, but the taller plants may need some support. Propagate by division in spring or fall or by root cuttings in early spring. Ensure there is a growth bud on the root cutting.

Tips

Use the taller species at the back of beds and borders. Inulas look good as specimen plants next to a pond, massed in a wild garden or spotted along the edge of a woodland. The flowers are great for cutting and can be used fresh and dried.

Recommended

I. ensifolia forms clumps of erect stems that bear narrow, mid-green foliage and spread by rhizomes. Golden yellow flowers bloom in mid- to late summer. Plants reach 30–60 cm tall and 30–46 cm wide.

I. helenium (elecampane) is a bold plant that has thick, rhizomatous roots and large, wide basal foliage to 90 cm long that gets smaller the higher it is on the sturdy stems. Plants grow 1.2–1.8 m tall and 90 cm–1.2 m wide in bloom but can grow larger in ideal conditions. The large, bright yellow flowers bloom in summer.

I. ensifolia

I. racemosa is a large plant, 1.8–2.4 m tall and 90 cm–1.5 m wide in bloom, with large, wide, coarse, green basal leaves 46–90 cm long that decrease in size as they rise up the stems. Large, spire-like clusters of light yellow flowers bloom from late summer to fall. Plants dry to a shiny bronze colour, providing winter interest. **'Sun Spear'** ('Sonnenspeer') bears bright yellow flowers. (Zones 4–8)

I. royleana (*I. macrocephala*) is an upright plant with green, 25 cm long leaves. Plants grow 46–60 cm tall and 30–46 cm wide. Very large yellow-orange flowers arise from black buds and bloom from mid-summer to early fall.

Problems & Pests

Plants are susceptible to powdery mildew.

I. royleana cultivar

Irish Moss

Pearlwort
Sagina

Height: 2.5–5 cm • **Spread:** 15–25 cm and more • **Flower colour:** white • **Blooms:** late spring to mid-summer • **Hardiness:** zones 3–7

Here's a little beauty that adds a low dimension to your garden. It is a great plant for softening the edges between beds and paved surfaces when allowed to creep onto the pavement. Although mostly grown for its awl-shaped foliage, Irish Moss' tiny, starry white flowers add to its charm. The plants might decline in hot, humid summers but will spring back in fall.

Planting

Seeding: start seeds in containers in a cold frame in fall
Transplanting: spring, after the risk of frost has passed
Spacing: 20–30 cm

Growing

Irish moss grows best in **full sun, partial shade** or **light shade** with protection from the hot afternoon sun in **moist, well-drained, neutral to slightly acidic** soil of **poor to moderate fertility**. This plant tolerates other soils but needs good drainage and moisture. 'Aurea' even tolerates clay soils. Plants suffer in drought and wet soils. These are tiny plants with shallow roots, so make sure they get enough water. Divide in spring or fall when plants are over-crowded or to propagate more plants.

Tips

Irish moss is a natural for alpine and rock gardens. Use it to edge beds, borders and containers, or spot plant and allow it to weave its way around perennials and shrubs. Irish moss can take some foot traffic and has been used as a lawn substitute, but it can't handle active use. It is a great plant for spaces and cracks in stone or brick walkways.

Recommended

S. subulata (Irish moss, awl-leaved pearlwort) forms mats of soft, dark, evergreen, mossy foliage. The stems can spread indefinitely and will root as they creep along the soil's surface. Tiny white flowers bloom from late

S. subulata (both photos)

spring to mid-summer. **'Aurea'** (Scotch moss, golden pearlwort) has bright chartreuse to yellow foliage.

Problems & Pests

Problems are infrequent, but crown rot can occur in poorly drained soils.

Use Irish and Scotch moss to create two-toned planting designs, such as a yin-yang symbol or a moderately sized chessboard or checkerboard.

Ironweed

Vernonia

Height: 60 cm–1.8 m • **Spread:** 60–90 cm • **Flower colour:** purple, magenta •
Blooms: late summer, fall • **Hardiness:** zones 3–8

Ironweed is a tough, attractive, low-maintenance plant that forms bushy,
upright clumps of sturdy stems and narrow, lance-shaped foliage. It lights up
the late-season border with its flat-topped clusters of bright magenta and
purple flowers. Ironweed combines and contrasts well with shorter, broad-
leaved plants.

Planting

Seeding: seeds require stratification; sow stratified seeds indoors or in containers in a cold frame in spring, or direct sow in fall
Transplanting: spring
Spacing: 60–90 cm

Growing

Grow ironweed in **full sun to partial shade** in **moist, well-drained** soil of **average fertility**. Deadhead to prevent self-seeding. Wind can carry the seeds well away from the plant. Staking may be required in windy locations. Divide the plant in early spring if necessary.

Tips

Ironweed looks great as a specimen or when mass planted. Use it at the back of beds and borders, around water features and in a wildflower or prairie garden. Ironweed also makes a good cut flower for fresh arrangements.

Recommended

V. arkansana (curlytop ironweed; *V. crinita*) **'Mammuth'** bears large clusters of bright violet flowers in late summer and fall on upright, unbranched or minimally branched stems. Plants grow 90 cm–1.8 m tall and 60–90 cm wide.

V. fasciculata forms bushy clumps of upright, unbranched stems with deep green foliage. It grows 60 cm–1.2 m tall and 60–90 cm wide and bears dense clusters of bright magenta flowers from late summer to early fall.

V. lettermanii (narrowleaf ironweed) **'Iron Butterfly'** bears very

V. noveboracensis (both photos)

narrow, light green foliage on branched stems. It grows 75–90 cm tall and wide. From late summer to early fall, it bears bright purple flowers. (Zones 4–8)

V. noveboracensis (New York ironweed) has branched stems with dark green foliage and grows 90 cm–1.8 m tall and 75–90 cm wide. It produces bright magenta flowers from late summer to mid-fall. (Zones 4–8)

Problems & Pests

Problems are infrequent, but plants may experience slugs, snails, leaf spot, rust and powdery mildew.

Jack-in-the-Pulpit

Indian Turnip
Arisaema

Height: 30–60 cm • **Spread:** 15 cm • **Flower colour:** purple, green •
Blooms: spring, early summer • **Hardiness:** zones 4–8

Jack-in-the-pulpit is sure to be a conversation starter. Each plant produces
only a few leaves and a single flower, and planting a group of them, or allowing them to form colonies, will make a definite impact in your garden. Eastern Canadian native *A. triphyllum* is likely the least bizarre-looking of all the
species listed below. Do not harvest these plants from the wild. Only purchase them through reputable plant dealers.

Planting

Seeding: start seeds in a cold frame in spring or fall; from seed, depending on species, expect two to four years to have a flowering sized plant; the size of the seed determines the amount of food reserves and, consequently, how large the plant will grow during the first season
Transplanting: spring or fall
Spacing: 15 cm for one plant and 30–46 cm if planting a large clump

Growing

Jack-in-the-pulpit grows well in **partial** or **light shade** and prefers **cool conditions**. The soil should be **fertile, loose, neutral to acidic, moist** and, especially, **well drained**. This plant requires shade in summer and regular watering until late summer, but it can be dry from late summer through the following spring. This tuberous perennial doesn't need to be divided, but the small offsets that

A. dracontium

grow at the base of the plant can be transplanted to propagate more plants. Plant the corms of most species 13–15 cm deep and dwarf species about 8 cm deep. Mulch the planting area for at least the first winter.

A. ringens

A. ringens

A. triphyllum

Tips

Jack-in-the-pulpit makes a great addition to the woodland garden. Include it in shaded borders, particularly those in moist spots on the north side of the house where the conditions are cooler.

Calcium oxalate crystals permeating all parts of this plant cause severe burning or death if ingested. Native Americans cooked and dried the tubers, nullifying these effects, and used them as a vegetable. They also used the tubers medicinally to treat cold symptoms, sore eyes and skin infections. The tubers have also been used for making laundry starch.

Recommended

A. dracontium (green dragon) produces one deeply lobed leaf and a narrow green flower in spring or early summer. It grows 30–90 cm tall.

A. ringens produces two lobed leaves and a single, large, green-and-purple-striped flower in early summer followed by showy clusters of red berries in fall. It grows about 30 cm tall. (Zones 6–8)

A. sikokianum produces two lobed leaves, sometimes with silvery variegations and a single, large, dark maroon-and-white-striped flower with a luminescent white, light bulb–like spadix (60 watts, at least!) in spring

A. triphyllum

followed by showy clusters of red berries in fall. It grows 30–46 cm tall. (Zones 5–8)

A. triphyllum produces one or two lobed leaves and a pale green or purple-striped flower in spring or early summer followed by showy clusters of red berries in fall. It grows 15–60 cm tall.

Problems & Pests
Slugs can cause trouble, and a few leaf diseases, such as a rust fungus, can occur.

Some Arisaema *are male, some are female, some are both and some change back and forth (paradioecious). Usually, those that do so are male when young, become female when they gather enough energy to have "babies," and, a year after giving birth (fruiting), often revert to being male.*

The flowers consist of a spathe and the spadix, which is held within the spathe.

Jupiter's Beard

Red Valerian
Centranthus

Height: 60–90 cm • **Spread:** 30–60 cm • **Flower colour:** red, pink, white • **Blooms:** summer • **Hardiness:** zones 4–8

Although not common at nurseries, Jupiter's beard is a light and airy plant. It was once very common in the cottage and cutting gardens of Europe after being introduced from its native Mediterranean region. It is a very aggressive self-seeder, to the point where care should be taken to cut back blooms as they fade. Doing so results in re-blooms that can carry into fall. Jupiter's beard makes a good contrast plant to darker coloured hardscaping features, such as walls and paver paths.

Planting

Seeding: start seeds in spring indoors or in a cold frame
Transplanting: spring or fall
Spacing: 30–46 cm

Growing

Jupiter's beard grows best in **full sun**. The soil should be of **average fertility, neutral to alkaline** and **well drained**. Too rich a soil or too much fertilizer encourages floppy, disease-prone growth. Division is rarely required but can be done in spring or fall to propagate desirable plants. Deadheading extends the blooming season and prevents excessive self-seeding.

If flowering slows during the summer months, cut the plant back by up to half to encourage new growth and more flowering. Cut the plant back in fall, leaving the basal foliage in place, to give it time to harden off before winter.

Tips

Jupiter's beard can be included in borders. It looks particularly impressive when left to self-seed in an unused corner of the garden, where it will form a sea of bright red flowers. The flowers are popular for fresh arrangements.

Recommended

C. ruber forms a bushy, upright plant. It bears large clusters of red, pink or white flowers on and off over the whole summer. Deadheading will encourage sporadic blooming through fall, but extending the bloom too long will shorten the life of the plant. **'Albus'** bears white flowers.

C. ruber 'Albus'

Jupiter's beard is rarely plagued by any pests or diseases.

These old-fashioned flowers were once common in cut-flower gardens.

C. ruber

Kalimeris

Japanese Aster
Kalimeris

Height: 30–60 cm • **Spread:** 30–60 cm • **Flower colour:** white or light blue with a yellow centre • **Blooms:** late spring to frost • **Hardiness:** zones 4–8

Tiny, white, almost baby's-breath-type flowers cloud the entirety of kalimeris, with just a bit of the small, pale green foliage showing underneath. More upright in seasons with average to low rainfall, the blossoms tend to flop with too much water or too much fertilizer. Planted in combination with flowering shrubs or other perennials, kalimeris can be considered a good team player in the plant world, working hard to make its companions look good.

Planting

Seeding: start seeds in containers in a cold frame in late fall or winter; seeds need a cold treatment to germinate
Transplanting: during cooler weather in spring or fall
Spacing: 46–60 cm

Growing

Kalimeris grows well in **full sun to partial shade** in **moderately fertile, moist, well-drained** soil, but it does well in most soils. It is quite weather and heat tolerant when established. This plant can be divided in spring or fall to rejuvenate it when it loses vigour or to propagate it. Shear plants back 15–25 cm after the first flush of flowers to rejuvenate the plant and to encourage more blooming.

Tips

Use kalimeris in beds, borders, informal gardens and at the edge of a woodland. It looks great when planted in large swaths or when dotted in the landscape as an accent plant. It also makes a good filler between colourful, coarse-textured plants and is a good companion to plants that are past their prime by mid-summer.

Recommended

K. incisa (cutleaf Japanese aster) **'Blue Star'** forms compact mounds of dark green foliage and produces yellow-centred, light blue, daisy-like flowers from early summer to early fall. Plants grow 30–46 cm tall and 30–60 cm wide. **'Variegata'** grows 46–60 cm tall and wide, with cream and green variegated foliage.

K. incisa 'Blue Star'

K. pinnatifida (Japanese aster) is an upright, bushy perennial that grows 46–60 cm tall and wide and spreads slowly to form large clumps. It has serrated, apple-green foliage and long-lasting, double, white flowers that have pale, buttery yellow centres. It blooms from late spring to frost.

Problems & Pests

Kalimeris rarely suffers from any pest or disease problems. It is more mildew resistant than many other asters.

K. pinnatifida

Kenilworth Ivy

Ivy-leaved Toadflax
Cymbalaria

Height: 5–8 cm • **Spread:** 30–60 cm • **Flower colour:** violet, white • **Blooms:** late spring to early fall • **Hardiness:** zones 4–8

This tough, low-maintenance plant is the same Kenilworth ivy that is grown as a houseplant. It is not a true ivy (*Hedera* spp.) but is in the same family as snapdragons and lobelia. The family resemblance shows up in the tiny flowers, but you need to be close to get a good look. Kenilworth ivy likes cool summers and dappled shade and suffers in high heat and humidity. It is worth trying in zone 3.

C. muralis (both photos)

Planting

Seeding: sow seeds directly in spring or fall or in containers in a cold frame in spring; do not cover the seeds because they need light to germinate
Transplanting: spring
Spacing: 30–60 cm

Growing

Grow Kenilworth ivy in **partial to light shade** in **well-drained** soil. It tolerates full sun as long as the soil remains moist and is happier when shaded from the hot afternoon sun. Kenilworth ivy needs regular watering during the first year to help the plants establish, with only minimal watering required after that. It will self-seed in good conditions. Propagate plants by division in spring, by stem cuttings and by layering. Cut back stems that are reaching out of bounds.

Tips

Kenilworth ivy looks great rambling through a woodland garden or cascading down a stone or rock wall. Use it in rock gardens or trailing over the edges of hanging baskets and containers. It can take light foot traffic and is used in the cracks of paver and flagstone paths and as a ground cover. Kenilworth ivy can also climb but may need a little guidance.

Recommended

C. muralis is a low-growing, trailing or climbing evergreen perennial with smooth, lobed, round to heart-shaped, glossy, dark, evergreen foliage. It grows 5–8 cm tall and the stems, which root at the nodes, trail 46–60 cm. Tiny, two-lipped, light violet flowers with two yellow dots in the centre and a short spur on the back bloom from late spring to early fall. The flowers are on long stalks that hold them above the foliage. **'Nana Alba'** is smaller than the species, with smaller leaves and stems that trail 30–46 cm. The flowers are white and also have the two yellow dots.

Problems & Pests

Kenilworth ivy rarely suffers any pest problems.

Knautia

Knautia

Height: 30–90 cm • **Spread:** 30–60 cm • **Flower colour:** dark red
Blooms: early summer to early fall • **Hardiness:** zones 4–8

Knautias are tough, easy-to-grow, long-blooming but somewhat short-lived perennials that self-seed enough to always maintain their presence in your garden. They are truly drought tolerant and have survived in my rain-sheltered foundation planting with minimal care, happily blooming all summer long. They have seeded themselves in the old fescue lawn and behind a euphorbia in one of the driest part of my landscape. As a bonus, the flowers attract butterflies to the garden.

Planting

Seeding: seeds need cold stratification; sow seeds directly in fall, in containers in a cold frame in late winter to early spring, or indoors after stratification
Transplanting: spring
Spacing: 30–60 cm

Growing

Knautia grows best in **full sun** in **moderately fertile, well-drained, neutral to alkaline** soil but is tolerant of a range of soils as long as they are well drained. The stems may require some support to prevent excessive flopping. Established plants are fairly drought tolerant but do best with infrequent, deep watering. Deadhead to encourage more blooms. Cut plants back to the ground in fall. Propagate by taking basal cuttings in spring.

Tips

Knautia looks good as a specimen or when mass planted in beds, borders and containers. It is a natural fit for wildflower and prairie plantings, and its somewhat lackadaisical habit works well in more informal settings. The flowers are long lasting as cut flowers.

Recommended

K. macedonica is bushy, well-branched and clump-forming, growing 60–90 cm tall and 30–60 cm wide. It bears grey-green to green foliage; the basal foliage goes from not lobed to somewhat lobed, and the stem leaves have deeper lobes. Double, dome-shaped maroon to

K. macedonica (both photos)

dark red flowers with white stamens bloom on upright to lax, wiry stems. **'Mars Midget'** is a smaller, sturdier selection that grows 30–46 cm tall and wide and bears maroon to ruby red flowers.

Problems & Pests

Plants may experience powdery mildew and aphids but are deer resistant.

Lady's Slipper Orchid

Cypripedium

Height: 15–75 cm • **Spread:** 25–41 cm • **Flower colour:** yellow and maroon, pink and white • **Blooms:** late spring to mid-summer • **Hardiness:** zones 2–7

Lady's slippers are found across Canada in moist forests, bogs and tallgrass prairies. They are long-lived plants once they are established. However, the plants need to form a symbiotic relationship with a mycorrhizal soil fungus for their survival, especially when the seeds are germinating, and can be hard to establish. Purchase only from reputable nurseries whose plants have been grown from seed or tissue culture.

Planting

Seeding: sow fresh, ripe seeds in containers filled with high-quality, fungally dominant compost or into the soil around existing plants; do not cover the seeds; germination and speed of growth to maturity are very slow

Transplanting: spring or fall

Spacing: 30–41 cm

Growing

Lady's slipper orchid grows best in **partial shade to full shade** in **loamy to sandy-loam, humus-rich, moist, well-drained, cool** soil. Plants will adapt to full sun as long as the soil is moist and there is shade from the hot afternoon sun. *C. calceolus* prefers neutral to alkaline soil, and *C. reginae* prefers slightly acidic conditions but adapts to a range of soil types. Plants benefit from a thick mulch of fallen leaves over winter. Topdress regularly around the plant with well-composted leaf mould. Gently divide established clumps in early spring or early fall and replant immediately. Be sure to include some of the existing soil because it contains the symbiotic mycorrhiza.

Tips

Lady's slippers make excellent specimen plants in rock and woodland gardens. They can also be used in a prairie planting where the tall grasses help keep the plants shaded. The stems and leaves of *C. calceolus* and *C. reginae* are densely covered with fuzzy hairs, which may cause dermatitis in some people.

Recommended

C. calceolus is a clump-forming plant with upright, unbranched stems. It grows 15–75 cm tall and 30–41 cm wide. The plant has light to medium green foliage that spreads slowly by creeping rhizomes. The unusual flowers have a yellow pouch (labellum) that often has purplish streaking and maroon-streaked, green to dark maroon petals and sepals. The plant blooms from late spring to mid-summer with 1–2 flowers per stem and produce a rose-like scent.

C. reginae (showy lady's slippers) is an upright species with medium to dark green foliage that bears large, fragrant flowers with white petals and a heavily pink-tinged, white labellum from late spring to early summer. Plants grow 25–75 cm tall and 25–30 cm wide.

Problems & Pests

Lady's slippers are prone to slugs, snails, rust, leaf spot and grey mould.

C. calceolus

Lemon Balm

Melissa

Height: 30–60 cm • **Spread:** 30–60 cm • **Flower colour:** white, light yellow, light purple • **Blooms:** summer • **Hardiness:** zones 3–7

Here is another herb that deserves to be used more in the perennial garden. The low mounds of wonderfully fragrant, useful foliage blend in well with flowering perennials, and the variegated and golden varieties provide a boost of bright colour in an otherwise green-foliaged location.

Planting

Seeding: sow seeds in containers in a cold frame in spring
Transplanting: spring
Spacing: 30–46 cm

Growing

Lemon balm prefers **full sun** but grows quite successfully in light, dappled shade. The ideal soil is **fertile, moist** and **well drained,** but this plant can tolerate poor, dry soils. Divide in spring or fall.

Cutting stems for use encourages dense, vigorous growth. Whole plants can be cut back in early summer to encourage dense growth. It's best to remove the flowers as they emerge, to reduce self-seeding.

M. officinalis (both photos)

The leaves can be harvested fresh or dried for teas, both hot and cold. They're also useful to flavour desserts and savoury dishes.

Tips

Herb gardens are often the preferred location for this useful perennial, but lemon balm also works well as a fragrant filler in containers, mixed beds and borders. Like the closely related mints, it may spread throughout your garden.

Recommended

M. officinalis is a bushy, dense-growing perennial with roughly textured, hairy leaves that are fragrant and flavourful when bruised or crushed. The inconspicuous light yellow to white flowers bloom in summer. **'All Gold'** has golden yellow foliage and light purple flowers. **'Aurea'** ('Variegata') bears dark green and golden yellow variegated foliage.

Problems & Pests

Lemon balm rarely suffers from pests.

Lilyturf

Liriope

Height: 20–46 cm • **Spread:** 46 cm • **Flower colour:** purple, purple-blue, white • **Blooms:** late summer through mid-fall • **Hardiness:** zones 4–8

Liriope is a popular groundcover in many states south of the border and is slowly gaining recognition here in Canada. Established plants are almost impervious to drought, heat, humidity and most garden pests and diseases. They require little care other than the occasional watering and a trimming in spring to make room for the new growth. Lilyturf is worth pushing the hardiness zone limits.

Planting

Seeding: difficult from seed but can self-sow in ideal conditions; seeds may or may not need sunlight to germinate, and cold stratification is recommended; try direct sowing ripe seeds in spring
Transplanting: spring
Spacing: 30–46 cm

Growing

Lilyturf grows best in **light** or **partial shade** but tolerates both full sun and full shade well. The soil should be of **average fertility, humus rich, acidic, moist** and **well drained**. A sheltered location is preferable because the ever-green leaves are prone to drying out in winter. Divide clumps in spring.

L. muscari, with *Lagerstroemia* in bed

L. muscari 'Pee Dee Gold Ingot'

A line trimmer, lawnmower or hedge shears can be used to cut back the spent foliage in late winter. A lawn-mower with a bagging attachment makes cleanup quick and easy. Be sure to do this before new growth emerges.

Tips

Lilyturf makes a fantastic, dense groundcover, ideal for keeping weeds down in a variety of locations. Include it in beds, borders, wood-land gardens and near water features. Also try lilyturf under trees where grass won't grow.

L. muscari 'Variegata'

L. muscari 'Pee Dee Gold Ingot'

L. muscari 'Monroe White'

Recommended

L. muscari forms low clumps of strap-shaped evergreen leaves. It grows about 20–46 cm tall, spreads about 46 cm and bears spikes of purple flowers from late summer through fall. **'Big Blue'** bears large spikes of purple-blue flowers. **'Monroe White'** bears white flowers. **'Pee Dee Gold Ingot'** has golden yellow to chartreuse leaves that are bright yellow when new. The flowers are light purple. **'Variegata'** has green-and-cream-striped leaves and bears purple flowers. 'Variegata' and 'Pee Dee Ingot' are hardy to zone 6. (Zones 5–8)

L. spicata (creeping lilyturf) grows 20–30 cm tall and is a rapidly spreading, potentially invasive species that does not form clumps. It is good for lawn replacement and erosion control. **'Silver Dragon'** does not grow as densely as most lilyturfs. It has slender, highly variegated green-and-white leaves and lavender flowers and stands about 30 cm tall.

Problems & Pests

Rare problems with root rot, anthracnose and slugs can occur.

Liriope was named after a Greek nymph.

Liverwort

Hepatica

Height: 8–15 cm • **Spread:** 10–15 cm • **Flower colour:** purple, blue, pink, white • **Blooms:** early to late spring • **Hardiness:** zones 4–8

Liverwort is a great early-spring bloomer and is definitely on the list of plants that are underused in the garden. The plants below are native to the eastern half of our country and are very low maintenance once established. Liverwort is well worth trying in zone 3. Please do not remove these plants from the wild. Purchase plants only from reputable nurseries and garden centres that grow stock from seed.

Planting

Seeding: seeds need cold stratification; sow fresh, ripe seeds directly or in containers in a cold frame in fall; stratified seeds can be sown indoors in spring

Transplanting: spring; some gardeners wait for the second year before planting out

Spacing: 10–15 cm

Growing

Liverwort grows best in **partial to full shade** in **moist, well-drained, neutral to alkaline** soil that has lots of good **compost** mixed in. Use leaf mould or high-quality, fungally dominant compost as topdressing after the plant has finished flowering. The plant can be divided in spring; however, liverwort resents root disturbance, and divisions may be slow to re-establish. Do not cut the foliage back until spring when new leaves start to appear. Fall cutting reduces the available nutrition from the leaves that the plant needs to survive.

Tips

Liverwort looks best when planted in random groups of 3–5 or more. Use liverwort for naturalizing shady spots and in woodland gardens, shady beds and borders.

Recommended

H. nobilis is a small, clump-forming plant that produces distinct, broadly oval, three-lobed, medium to dark green foliage that is shiny on the top and densely hairy on the underside. The lobes are deeply cut. The foliage turns rusty red to burgundy purple in winter. The stems and emerging flowers are also densely hairy. Long-lasting, star-shaped, white, blue, bluish purple, purple or pink flowers bloom from early to late spring. The leaves of **var. *acuta*** (sharp-leaved liverwort; *H. acutiloba*) have pointed lobes. **Var. *obtusa*** (round-leaved liverwort; *H. nobilis*) is an earlier flowering and slightly hardier variety than var. *acuta* and has the same great foliage, but the lobe ends are rounded. These two varieties hybridize freely, producing a range of leaf shapes.

Problems & Pests

Occasional problems occur from slugs, snails, rust and smut.

Liverwort was once seen as the cure-all for most liver disorders. Aboriginal peoples used liverwort tea for relief from coughs and sore throats.

H. nobilis var. *acuta*

Mayapple

Podophyllum

Height: 15–75 cm · **Spread:** 30–90 cm· **Flower colour:** red, purple, white, pink · **Blooms:** spring, summer · **Hardiness:** zones 5–8

P. peltatum is an Eastern Canadian native that makes a handsome, vigorous woodland groundcover, but do not ignore the non-native species and hybrids that make fabulous specimen plants. Make sure the soil has a lot of humus for the best results; add high-quality, fungally dominant compost or leaf mould to the soil, or use mulch.

Planting

Seeding: start seeds in a cold frame in late summer or fall
Transplanting: spring
Spacing: 30–60 cm

Growing

Mayapples grow well in **partial** or **full shade**. The soil should be **fertile, humus rich** and **moist**. Divide in spring or late summer to propagate more plants. Use mulch in fall to protect the plants from freeze-thaw cycles. In woodland gardens, allow the fallen leaves to serve this purpose.

P. 'Kaleidoscope'

Tips

These plants are naturally at home in a moist woodland garden and make excellent additions to a pondside planting. Include them in shaded borders, and keep the soil mulched to hold in moisture.

Mayapple is all **poisonous**—leaves, stems, roots and fruit, except for the **fully ripened** fruit. Eating unripe fruit can be cathartic. The toxicity helps protect the fruit until seeds mature. Indian apple, ground lemon, raccoon berry and hog apple are some other common names of the native species.

P. peltatum with English ivy

P. 'Kaleidoscope'

P. peltatum

Recommended

P. delavayi is a clump-forming plant that grows 15–46 cm tall with a spread of about 46 cm. The velvety green foliage is mottled with red, bronze and purple. Each leaf has 5–8 lobes, with each lobe divided into three more lobes. Pinkish red flowers in summer are followed by apricot-coloured fruit.

P. hexandrum (Himalayan mayapple) forms a clump of green and purple splattered leaves. It grows about 46 cm tall and spreads about 30 cm. Flowers are brilliant pink and are followed by bright red fruit.

P. 'Kaleidoscope' is a dazzling hybrid with hexagonal, umbrella-like, star-shaped leaves that reach 46 cm and are patterned like a kaleidoscope in green, splashed with silver, cream, purple-black and bronze. It grows about 60 cm tall, with an equal spread. Clusters of up to 20 large, wine red flowers, about 5 cm across, hang just below the leaf in summer.

P. peltatum (mayapple, American mandrake) forms a low-spreading mass of large glossy umbrella-like leaves, sometimes so dense that the earth beneath them remains dry even during a torrential rainfall. It grows about 46 cm tall and spreads up to 1.2 m. Fragrant white or light pink flowers are borne in spring. The fruit starts off green and turns yellow.

P. pleianthum

P. peltatum

P. pleianthum (Chinese mayapple) forms an upright clump 45–75 cm tall and spreads up to 90 cm. The star-shaped leaves are glossy green, and the summer flowers are deep red or purple, followed by silvery fruit.

Problems & Pests
Slugs may attack young foliage.

Mondo Grass

Ophiopogon

Height: 10–30 cm · **Spread:** 20–30 cm · **Flower colour:** white, pale purple · **Blooms:** summer · **Hardiness:** zones 6–8

Mondo grass is a close cousin of lilyturf and is the only zone 6 listed plant in this book. It is definitely worth trying in zone 5, especially with a warm microclimate. This plant is great for growing in containers or troughs. Containers can be placed in a warmer location, such as an unheated garage, over the winter. The black varieties certainly add a different element to a landscape design and are excellent contrast plants for creating standout displays with yellow-, blue- or silver-leaved plants.

Planting
Seeding: start seeds in a cold frame in fall
Transplanting: spring or fall
Spacing: 20–30 cm

Growing
Mondo grass grows well in **full sun** or **partial shade** in **fertile, humus-rich, slightly acidic, moist, well-drained** soil. Divide plants in spring, just as they begin to sprout new leaves. Use a mulch of evergreen branches to provide the plants with winter protection without smothering them or causing rot.

Tips
Mondo grass makes an excellent groundcover and edging plant for beds and borders. Plant it in containers placed around decks and pools. Try it in a rock garden and dotted among your spring bulbs.

Recommended
O. japonicus forms low grass-like clumps of dark green foliage. It bears clusters of white flowers in summer. It grows 20–30 cm tall and spreads about 30 cm. **'Compactus'** is a wee plant at 5 cm high. **'Kyoto Dwarf'** only grows 10 cm tall. **'Nanus'** (dwarf mondo grass) grows 10–20 cm tall with a spread of about 20 cm.

O. planiscapus forms a compact clump of strap-shaped, dark green leaves. It grows 10–20 cm tall with a spread of about 30 cm and bears pale purple flowers in summer. **'Nigrescens'** ('Black Dragon,' 'Nigra'; black mondo grass) has striking, dark—almost black—foliage. **'Black Night,'**

O. japonicus

'Ebony Knight' and **'Ebony Night'** are other available cultivars, 10–15 cm tall, and are possibly all the same plant.

Problems & Pests
Slugs may feed on new leaves in spring.

O. planiscapus 'Nigrescens'

Mountain Mint

Pycnanthemum

Height: 46–90 cm • **Spread:** 30–60 cm • **Flower colour:** white, pink •
Blooms: summer, early fall • **Hardiness:** zones 4–8

Mountain mints are low-maintenance members of the mint family. It is easy
to identify the mint family—their stems are distinctly four-sided, and the
plants are often aromatic. The two mountain mints mentioned below emit a
strong, minty aroma when bruised, crushed or brushed up against. The leaves
can be used to flavour teas. Mountain mints are deer resistant, and the plants
spread by rhizomes. *P. muticum* can spread aggressively, but *P. virginianum*
spreads slowly. Mountain mints are worth trying in zone 3 with protection.

Planting

Seeding: direct sow seeds in spring or fall
Transplanting: spring
Spacing: 30–60 cm

Growing

Mountain mints grow best in **full sun, partial shade** or **light shade** in **moist, well-drained, fertile** soil. Allow the soil to dry out some between watering. Mountain mints will adapt to a drier soil and a range of soil textures. Divide the plants in spring. Shear back *P. muticum* in early summer for really bushy plants.

Tips

Use mountain mints in herb gardens, beds, borders, meadow plantings, wildflower gardens, woodland gardens and near water features such as ponds and creeks. They are great for naturalizing, and *P. muticum* looks especially good when planted en masse where the silvery bracts really stand out. They make good cut flowers for fresh and dried arrangements and are good candidates for container plantings.

Recommended

P. muticum (short-toothed mountain mint, big leaf mountain mint) is a vigorous, upright, bushy plant that grows 46–90 cm tall and 46–60 cm wide. It has oval, mid- to dark green foliage with small-toothed margins and showy, silvery white bracts. Tight clusters of light pink flowers bloom throughout summer into early fall. The flower clusters of the *Pycnanthemum* genus bloom from

P. muticum (both photos)

the outside of the cluster toward the centre.

P. virginianum (Virginia mountain mint, wild basil, prairie hyssop) is an erect, well-branched, bushy plant, 46–90 cm tall and 30–46 cm wide, with narrow, pointed, mid- to dark green foliage. It bears many flattened clusters of small, white, two-lipped flowers that are sometimes dotted with purple bloom in mid- to late summer.

Problems & Pests

Mountain mints have no serious pest problems and are deer resistant. Plants under stress may experience rust.

Northern Maidenhair Fern

Adiantum

Height: 30–60 cm • **Spread:** 30–60 cm • **Flower colour:** grown for foliage • **Blooms:** spores produced in summer • **Hardiness:** zones 3–8

Northern maidenhair fern has a horizontal, layered form that adds wonderful texture to the shade garden. The delicate, nearly lime green foliage and little black wire-like stems provide movement in the garden. This fern sends up its fiddleheads a few weeks later than other ferns, so don't think it has died.

Planting

Seeding: sow ripe spores outdoors in early fall
Transplanting: spring
Spacing: 20 cm

Growing

Northern maidenhair fern grows best in **partial to full shade,** in **slightly acidic, fertile, moist, well-drained** soil.

Tips

Northern maidenhair fern works best at the edge of a woodland garden. It makes a good addition to a shaded border or shaded rock garden and does well in a streamside planting. When left to its own devices, maidenhair fern spreads to form colonies.

This fern is easy to propagate in fall. Just slice off a section of the thick root mass and replant it in a cool spot.

Recommended

A. pedatum is an upright, deciduous plant with dark brown to black stems.

A. pedatum (both photos)

It bears branched, horizontally oriented, lance-shaped fronds. The lobed, fan-shaped, green foliage turns yellow-green to yellow in fall. A waxy coating on the leaflets rapidly sheds water and raindrops.

Problems & Pests

Northern maidenhair fern rarely suffers from any problems.

Some Adiantum *species have been used to cure bronchitis, coughs and asthma. They are also known as good hair tonics and restoratives.*

Ox-Eye Daisy

Buphthalmum

Height: 46–60 cm • **Spread:** 46–60 cm • **Flower colour:** yellow • **Blooms:** early summer to early fall • **Hardiness:** zones 3–7

Ox-eye daisy is an underused and hardy perennial found on slopes, meadows and open woodlands in Europe. It is a low-maintenance, adaptable plant that blends well with other plants in most settings. It has attractive foliage and wonderful daisy flowers that bloom for an extended period.

Planting

Seeding: sow seeds in containers in a cold frame in spring or fall, or indoors in spring
Transplanting: spring
Spacing: 46 cm

Growing

Ox-eye daisy grows best in **full sun** in **moist, well-drained, slightly alkaline** soil. This plant grows reasonably well in partial shade and will adapt to most soils as long as they are well drained. Divide in early spring before the new growth appears.

Tips

Ox-eye daisy looks good when mass planted. Use it in mixed or perennial beds and borders, in wildflower plantings, by ponds and other water features, and in the cut-flower garden. Ox-eye daisy will attract butterflies to your garden.

Recommended

B. salicifolium forms 20–30 cm tall clumps of narrow, willow-like, dark green foliage. Clusters of large, deep yellow flowers with deeper yellow centres are held above the foliage on slender, upright stems. **'Alpengold'** bears bright, golden yellow blooms that are larger than the species.

Problems & Pests

Ox-eye daisy rarely suffers from any problems.

Ox-eye daisy flowers are great for adding to fresh flower arrangements and bouquets.

B. salicifolium

Patrinia

Patrinia

Height: 30 cm–1.8 m • **Spread:** 25–60 cm • **Flower colour:** yellow, white •
Blooms: mid-summer, late summer, fall • **Hardiness:** zones 5–8

The clear yellow blooms of patrinia rise to 1.5 m in thick, airy clusters and
have a pleasant fragrance. As the flowers fade, the chartreuse bracts become
showy. Patrinia's first full year is commonly spent establishing itself without
its showy blooms, but afterward, it is long lived and free flowering. The blooms
are good cut flowers. The foliage remains basal and turns reddish in fall. These
plants look good at the back of a border with purple to purple-blue flowered
plants, such as perennial salvia, in the front. Zone 4 gardeners might successfully grow patrinia.

Planting
Seeding: sow fresh, ripe seeds in containers in a cold frame in fall
Transplanting: spring
Spacing: 46–60 cm

Growing
Patrinia prefers to grow in **full sun to partial shade** in **moderately fertile, humus-rich, moist, well-drained** soil. *P. triloba* does best in morning sun and afternoon shade. Patrinia tolerates heat, humidity, rain, wind and cold, and it is somewhat drought tolerant. It will self-seed unless deadheaded. Divide in spring.

Tips
Patrinia can be used in beds and borders and at the edge of a woodland garden. It is a good filler in bouquets and arrangements and when mixed with bolder plants in the garden. The smaller species can be used as groundcovers and in rock gardens.

Recommended
P. scabiosifolia (golden lace) is an upright perennial, 90 cm–1.8 m tall and 46–60 cm wide. It has large, ferny, coarsely toothed, green foliage that may turn yellow or red-tinged in fall. Over a long period from late summer to fall, it bears airy, flat-topped clusters of small, fragrant, bright yellow flowers on tall, branched stems. **'Nagoya'** is a dwarf specimen that grows 90 cm tall.

P. triloba is a vigorous, clump-forming perennial that spreads by

P. villosa

short stolons and grows 30–60 cm tall and 25–36 cm wide. It has red-tinged stems and deeply lobed, coarsely toothed, green foliage that turns yellow in fall. The airy clusters of fragrant, golden yellow flowers bloom from mid- to late summer.

P. villosa (white patrinia) is a fast-growing, clump-forming perennial that grows 60–90 cm tall and 46–60 cm wide. It has bright green, slightly hairy foliage and clusters of fragrant, white flowers from late summer to fall.

Problems & Pests
Rust, snails and slugs may be problematic for patrinia.

P. scabiosifolia

Perennial Salvia

Sage
Salvia

Height: 30 cm–1.2 m • **Spread:** 30–90 cm • **Flower colour:** cream, purple, blue, pink • **Blooms:** late spring, summer, early fall • **Hardiness:** zones 3–8

Spiky salvias combine beautifully with virtually all perennials with a mounding habit. Don't confuse perennial salvia with annual salvias sold as bedding plants. They are from the same genus but are different species. Perennial salvias are reliable and quite hardy, returning in a larger clump each year, provided the plants are grown in ample sun and soil that drains well. The dwarf cultivar 'May Night' was the Perennial Plant of the Year in 1997.

Planting

Seeding: species can be started in early spring; cultivars do not come true to type
Transplanting: spring
Spacing: 46–60 cm

Growing

Salvias prefer **full sun** but tolerate light shade. The soil should be of **average fertility** and **well drained**. These plants benefit from a light mulch of compost each year. They are drought tolerant once established. Division can be done in spring, but the plants are slow to re-establish and resent having their roots disturbed. They are easily propagated by tip cuttings.

S. officinalis 'Tricolour' with cilantro, basil and chives

S. officinalis 'Tricolour'

S. officinalis 'Icterina'

Deadhead to prolong blooming. Trim plants back in spring to encourage new growth and keep plants tidy. New shoots sprout from old, woody growth.

Tips

All *Salvia* species are attractive plants for the border. Taller species and cultivars add volume to the middle or back of the border, and the smaller specimens make an attractive edging or feature near the front. Perennial salvias can also be grown in mixed planters.

S. nemorosa 'East Friesland'

S. officinalis 'Purpurescens'

S. officinalis

Recommended

S. azurea var. grandiflora (azure sage) is an openly branched, upright plant that grows 90–120 cm tall and 60–90 cm wide. It produces azure blue blooms in late summer and into fall. (Zones 5–8)

S. nemorosa (*S.* x *superba;* perennial salvia) is a clump-forming, branching plant with grey-green leaves. It grows 46–90 cm tall and spreads 46–60 cm. Spikes of blue or purple flowers are produced in summer. **'Blue Queen'** bears dark purple-blue flowers. **'Lubeca'** bears long-lasting purple flowers. (Zones 3–7)

S. officinalis

S. pratensis

S. officinalis (common sage) is a woody, mounding plant with soft grey-green leaves. It grows 30–60 cm tall and spreads 46–90 cm. Spikes of light purple flowers appear in early and mid-summer. **'Berggarten'** ('Bergarden') has silvery leaves about the size and shape of a quarter. **'Icterina'** ('Aurea') has yellow-margined foliage. **'Purpurea'** has purple stems. The new foliage emerges purple and matures to purple-green. **'Tricolour'** has green or purple-green foliage outlined in cream. New growth emerges pinkish purple. It is the least hardy of the variegated sages. (Zones 4–7)

S. pratensis (meadow sage, meadow clary) sends up lavender blue flower stems to 60–90 cm tall and 46–60 cm wide from a foliage clump of oblong, crinkled leaves. It flowers from mid-summer through fall.

S. sclarea (clary sage) is a short-lived perennial or biennial that grows up to 90 cm tall and spreads about 30 cm. It forms a mound of large, fuzzy leaves and bears large, loose spikes of purple-bracted cream, purple, pink or blue flowers in late spring and early summer. (Zones 4–8)

S. x sylvestris grows 75–90 cm tall and about 30 cm in spread with greyish green foliage and deep violet blue flower spikes that are long lasting. It is often confused with the very similar *S. nemorosa*. Cultivars have been listed under both species at different times. **'Blue Hill'** grows 46–60 cm tall with numerous flower spikes of clear blue flowers. **'May Night'** grows

S. officinalis 'Icterina' with Italian parsley, thyme and tarragon

to 46 cm and bears large, deep purple-blue flowers. **'Rose Queen'** bears unique rosy purple flowers, but the growth is somewhat floppier than that of the other cultivars.

S. verticillata 'Purple Rain' is a low, mounding plant with crinkled foliage that bears light red to violet blooms beginning in early summer and lasting until early fall. The colourful bracts remain long after the flowers fade. It grows 46 cm tall with an equal spread. (Zones 5–8)

Problems & Pests

Scale insects, whiteflies and root rot (in wet soils) are the most likely problems.

Common sage has aromatic foliage that is used for flavouring in many dishes.

The genus name Salvia *comes from the Latin* salvus, *"save," referring to the medicinal properties of several species.*

Perennial Sunflower

Helianthus

Height: 90 cm–2.4 m • **Spread:** 60–90 cm • **Flower colour:** yellow with brown to purple centres • **Blooms:** late summer, fall • **Hardiness:** zones 4–8

Perennial sunflowers brighten up the late-summer and early-fall garden when not much else is in fresh bloom. The main attractions are their tremendous height and late-season bloom time. *H. salicifolius* is a finer texture than the other species listed, and it adds an interesting texture to the garden. Grow Jerusalem artichokes for the flowers and the tasty, edible tubers.

Planting
Seeding: start in spring; seedlings may not come true to type
Transplanting: spring
Spacing: 60–90 cm

Growing
Sunflowers grow best in **full sun.** They flower less and become lanky in shaded conditions. Ideally, the soil should be of **average fertility, neutral to alkaline, moist** and **well drained.** *H.* x *multiflorus* prefers a constantly moist soil, while *H. salicifolius* is drought tolerant. If powdery mildew is a problem on the lower foliage, pull out a few stems to increase airflow. Be careful with the fertilizer—more than a light application in spring can lead to excessive leaf growth and diminished flowering.

H. decapetalus 'Plenus'

Plants should be cut back hard after flowering. They can also be cut back in early summer to produce shorter but later-flowering plants that are more floriferous. Divide every three or so years in spring or fall.

H. x *multiflorus* 'Loddon Gold'

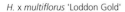

H. decapetalus 'Plenus'

Tips

These impressive perennials deserve a spot at the back of the border or in the centre of an island bed. Sunflowers are tall plants that can provide a quick privacy screen in an exposed garden. Water-loving *H.* x *multiflorus* can also be planted near a pond or other water feature. *H. salicifolius* can be used in dry and underwatered areas of the garden.

These plants are in the same genus as the annual sunflowers (H. annuus), but the perennial versions have smaller, more plentiful flowers. Butterflies and bees are attracted to these plants.

Recommended

H. decapetalus (thin-leaf sunflower) is a rhizomatous perennial that forms a large, upright clump 90 cm–1.5 m tall and spreads about 90 cm. Many daisy-like flowers with yellow petals and brownish yellow centres are borne in late summer and fall. **'Plenus'** ('Flore-Pleno') bears large, double, golden yellow flowers with pointed petals.

H. **'Lemon Queen'** has single, bright but soft lemon yellow blooms on 1.8–2.4 m tall, sturdy, upright stems. The spread is about 90 cm. It blooms from mid-summer to frost.

H. x *multiflorus* (many-flowered sunflower) forms a large, upright clump. It grows 90 cm–1.8 m tall and spreads about 90 cm. It features daisy-like flowers borne in late summer and

H. salicifolius

fall, with yellow petals and brown centres. **'Capenoch Star'** forms a large clump of stems with narrow leaves. It grows 1.2–1.5 m tall and bears light yellow flowers from mid-summer to early fall. **'Loddon Gold'** bears golden yellow, double flowers. **'Soleil d'Or'** bears double, yellow flowers.

H. salicifolius (willow-leaved sunflower) is a large, clump-forming, rhizomatous plant with narrow leaves. It grows 90 cm–1.8 m tall, occasionally to 2.4 m tall, and spreads 90 cm. It bears daisy-like, yellow flowers with purple-brown centres in fall.

H. tuberosus (Jerusalem artichoke, sunchoke) is a tall, bold plant that reproduces by seed and by fleshy rhizomes. The rhizomes bear white, red or purple, edible, knotty tubers. Plants grow 1.8–2.4 m tall and 60–90 cm wide. The flowers are light to bright yellow with yellow centres and bloom from mid-summer to fall.

Problems & Pests

Problems with powdery mildew, downy mildew, fungal leaf spot and leaf-chewing insects such as caterpillars, beetles and weevils can occur.

Little is known about the sunflower's earliest history, but the conquistadors found many pure gold representations of the sunflower embellishing Aztec and Incan temples.

The tubers of Jerusalem artichoke can be eaten fresh or cooked as one would prepare potatoes. Leave the tubers in the ground until you are ready to use them. Tubers subjected to frost become a little sweeter to the taste.

Pitcher Plant

Sarracenia

Height: 15–90 cm · **Spread:** 15–90 cm · **Flower colour:** purple, red, pink, purplish red, yellow · **Blooms:** spring · **Hardiness:** zones 2–8

How cool are bug-eating plants, especially when one is a native species that is very hardy? *S. purpurea* is found across the country and is the floral emblem of Newfoundland and Labrador. If you have a boggy area, you really need to plant this species. If you live in the warmer zones and have a boggy area, your choices expand greatly. Boggy, acidic soil allows pitcher plants to thrive and also attracts moss, which likes to grow in the same conditions. Adding 25% to 50% peat moss to a poorly drained area will create a good area for pitcher plants.

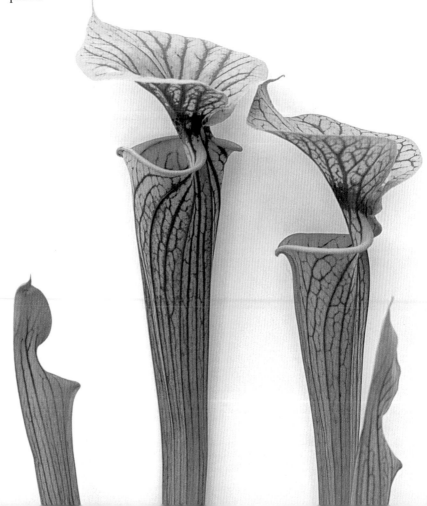

Planting
Seeding: seeds need a short period of cold stratification to germinate; sow stratified seeds into warm soil in the garden in spring
Transplanting: spring
Spacing: 15–60 cm

Growing
Pitcher plants prefer locations with **full sun to partial shade**. The soil should be **humus rich,** consistently **moist to boggy** and **acidic.** Pitcher plants thrive in nutrient-depleted soils, so fertilizing isn't necessary and may actually kill the plant. These plants absorb the necessary nutrients from the insects they consume.

S. flava

Division is required only if you want to propagate the plants. Carefully separate the crowns in mid-spring, avoiding damage to the roots, and replant the divisions immediately. Remove only dead foliage in fall—do not disturb the roots.

Tips
Pitcher plants are excellent for locations that are kept consistently moist but not flooded. The insect-eating tubes that rise out of the damp ground are certain to be a conversation piece. Do not dig these plants up from the wilderness. Buy them only from reputable nurseries.

Recommended
S. flava (yellow pitcher plant) produces erect, pitcher-shaped foliage in yellowish green with crimson veining. It bears large, bright yellow, pendulous flowers that often rise above the pitchers. The species grows 60–90 cm tall and wide. (Zones 6–8)

S. purpurea (common pitcher plant, huntsman's cup) forms a clump of prostrate to erect, jug-shaped, purple-veined, green, purple or reddish pitchers that grow 15–30 cm tall and wide. Nodding flowers in shades of purple, red, pink or purplish red appear in spring.

Problems & Pests
Scale insects, mealybugs and aphids can cause problems.

S. purpurea

Prairie Coneflower

Mexican Hat
Ratibida

Height: 60 cm–1 m • **Spread:** 30–60 cm • **Flower colour:** yellow, maroon, dark red •
Blooms: early summer to early fall • **Hardiness:** zones 3–8

Prairie coneflower is fast growing and hardy, has a long bloom period and
often blooms the first year if grown from seed. Native to southeastern BC,
this wonderful perennial has naturalized and spread all the way to Ontario; it
also can be seen in ditches and disturbed areas through the Prairies. Plus, you
can brew a pleasant-tasting tea from the leaves and flowers.

Planting

Seeding: seeds germinate best with cold stratification; sow seeds in containers in a cold frame in late winter to early spring or fall, or direct sow outdoors in fall
Transplanting: spring
Spacing: 30–46 cm

Growing

Prairie coneflower does best in **full sun** in **dry to average, well-drained, neutral to alkaline, moderately fertile** soil. Plants tolerate short periods of drought and soils of lower fertility and clay soils. Established plants are drought tolerant but may need additional water in fall and spring if the growing season was exceptionally dry. The bloom time may extend into fall with adequate moisture. Divide young plants in spring. Dividing plants may be difficult because of the long taproot.

Tips

Prairie coneflower is a natural for wildflower, meadow and prairie plantings. Use it as a specimen, or mass plant it in beds and borders. It makes a good cut flower, is worth trying in containers and is a great choice for xeriscaping. The flowers will attract bees, butterflies and beneficial predatory insects.

Recommended

R. columnifera is an upright perennial with hairy, grey-green, deeply lobed foliage. The plant develops a long taproot from which the stems arise. Long, thin, single or branched stems display the interesting flowers

R. columnifera (both photos)

at their tips. The flowers have reflexed, yellow petals and a black to brown, cone-shaped central disk. The flower petals (which are actually ray flowers) may be solid dark red to maroon, or dark red or maroon with yellow.

Problems & Pests

Rare problems with powdery mildew, downy mildew, smut and fungal leaf spots can occur.

Prairie Poppy Mallow

Wine Cups, Buffalo Poppy
Callirhoe

Height: 15–30 cm • **Spread:** 30–90 cm • **Flower colour:** magenta, purple, purple-red, purple-pink • **Blooms:** late spring to late summer • **Hardiness:** zones 4–8

If you have a hot and dry site, this plant is for you! Prairie poppy mallow is an easy-to-grow, low-maintenance, multiple-use perennial with bright flowers that bloom over a long period. It is finally starting to get the recognition it deserves, and the plants are becoming more available in nurseries and garden centres. It is easy enough to grow from seed if you can't find a plant locally. Prairie poppy mallow is well worth trying in zone 3.

Planting

Seeding: direct sow seeds in fall or early spring; stratification and scarification will aid germination
Transplanting: spring
Spacing: 30–75 cm

Growing

Prairie poppy mallow grows best in **full sun** in **well-drained, sandy** soil. It grows fairly well in partial shade and will adapt to many different soil types, including clay and moist soils, as long as the soils are well drained. Keep the plant dry in winter. Do not damage the taproot when transplanting. Propagate by taking softwood cuttings or basal cuttings in early summer. Division is possible but is very difficult because of the long taproot. Prairie poppy mallow tolerates salt spray when planted near roadways. The plant self-seeds but not aggressively; deadhead to remove this possibility.

Tips

Prairie poppy mallow looks great when mass planted or alone as a specimen. Use it in beds, borders, wildflower gardens, meadow plantings, rock gardens, containers and as edging. Prairie poppy mallow makes a good groundcover when spaced tightly together. The flowers are a good nectar source and will attract bees and butterflies to your garden.

Recommended

C. involucrata is a dense, low-growing, spreading, semi-evergreen perennial with procumbent or sprawling, hairy stems that arise from a large taproot. The fern-like foliage can be light to medium green or greyish green, and it has deep palmate lobes. Cup-shaped, brightly coloured flowers are held above the foliage on thin stems.

Problems & Pests

Aphids, spider mites, rust, powdery mildew, slugs and snails may cause problems. Crown rot can occur in poorly drained soils.

The long taproot and foliage of prairie poppy mallow are edible when cooked. The root is said to taste like sweet potatoes, and the leaves are great for use in soups and stews.

C. involucrata

Purple Moor Grass

Molinia

Height: 30 cm–2 m • **Spread:** 30–90 cm • **Flower colour:** reddish brown to purple • **Blooms:** mid-summer to fall • **Hardiness:** zones 4–8

Purple moor grass is a beautiful, easy-to-grow, very underused perennial grass. It offers great foliage with good fall colour and showy spikes of flowers on slender, arching stems that sway gently in a breeze. It should do well from Ontario eastward and on the West Coast. Prairie gardeners will need to keep an eye on the soil moisture. Purple moor grass is worth trying in zone 3.

Planting

Seeding: sow seeds in containers in a cold frame in spring
Transplanting: spring
Spacing: 30–90 cm

Growing

Purple moor grass grows best in **full sun to partial shade** in **moist, well-drained, neutral to slightly acidic, average to fertile** soil. In hot, dry areas, partial shade is best. Plants grow well in sandy soil and tolerate soils of low fertility, alkaline soils and clay soils as long as they are well drained. Ensure the plants receive adequate water in the dry climates. Divide in spring. Purple moor grass is slow to establish.

Tips

Purple moor grass looks great as a specimen plant and is stunning when massed. Use it in moist beds, borders, containers, at the edge of a woodland garden or near water features such as ponds and bogs. It is a fair groundcover when densely planted, and the cut flowers are good for fresh and dried arrangements.

Recommended

M. caerulea is a dense, tufted grass with long, thin to medium-width blades. The green foliage is purplish at the base and turns a nice yellow in fall. It grows 30–60 cm tall, and the clumps spread up to 75 cm wide. Yellow stems up to 90 cm tall bear spikes of reddish brown to purple flowers from mid-summer to frost. **'Moorhexe'** is an erect, compact selection with foliage that grows

M. caerulea subsp. *arundinacea* 'Skyracer'

30–46 cm tall and has yellow to orange fall colour. **Subsp. *arundinacea* 'Sky Racer'** reaches 2 m in bloom. The long, graceful, blue-green foliage grows 60–90 cm tall and to 90 cm wide and turns golden yellow in fall. **'Variegata'** bears green and creamy white-striped foliage that grows 46–60 cm tall and to 60 cm wide. Flowering stems appear a little later than with the species.

Problems & Pests

Purple moor grass rarely experiences problems. Even deer resist browsing.

M. caerulea 'Variegata'

Pussy Toes

Antennaria

Height: 2.5–30 cm • **Spread:** 20–46 cm • **Flower colour:** white, light pink, rose pink • **Blooms:** late spring to early summer • **Hardiness:** zones 1–8

Pussy toes are tough, mat-forming, native plants that spread slowly by stolons. They have rosettes of woolly, semi-evergreen to evergreen foliage and clusters of everlasting flowers that resemble the pads on a cat's paw. *A. rosea* is common and widespread across Canada and is found in a large variety of climatic and geographic conditions, so it will probably perform wonderfully in your landscape.

Planting

Seeding: germination may be slow and somewhat variable; cold stratification aids germination; sow seeds directly or in a cold frame in spring or fall

Transplanting: spring or fall

Spacing: 30–46 cm

Growing

Pussy toes grow best in **full sun to partial shade** in **well-drained** soil of **average to poor fertility.** They do well in moist conditions but prefer soil that is on the dry side. Plants perform admirably in sandy and gritty soils and are very drought tolerant when established. Divide in spring or fall.

Tips

Pussy toes look great in alpine and rock gardens, as groundcovers and as edging for beds and borders. These plants can take light foot traffic and are good for filling in the cracks and spaces in paver and natural stone paths. The cut flowers make lovely fresh or dried arrangements. These plants are an excellent choice for xeriscaping.

Recommended

A. alpina (alpine pussy toes) forms mats of silvery grey-green foliage that grows 2.5–5 cm tall and spreads 30–46 cm. Clusters of creamy white flowers are borne on flowering stems 10–25 cm tall.

A. dioica (stoloniferous pussy toes, mountain everlasting) has dense, spoon-shaped, grey-green foliage with silvery white hair on the

A. rosea

undersides. Plants reach 2.5–5 cm in height and 20–41 cm in width. Clusters of white or light pink flowers bloom on 15–25 cm tall stems. (Zones 3–8)

A. rosea bears clusters of rose-pink flowers on 10–30 cm tall stems. The grey-green foliage grows about 5 cm tall and is more upright than the other species listed here. The low, mat-forming clumps spread 20–30 cm. *A. rosea* is actually a complex of four distinct subspecies.

Problems & Pests

Pussy toes may experience slugs, smut, powdery mildew and leaf spot.

A. dioica

Ribbon Grass

Gardener's Garters, Reed Canary Grass
Phalaris

Height: 60 cm–1.2 m • **Spread:** 90 cm to indefinite • **Flower colour:** pale pink to white; grown for foliage • **Blooms:** early summer • **Hardiness:** zones 4–8

Why recommend a plant that can be difficult to keep in bounds? For wet areas, drier sunny slopes and challenging areas that require erosion protection, the rhizomes of ribbon grass spread quickly to form a dense weave that holds the soil in place. For its invasive qualities, plant the species. If you want pretty foliage and slightly less aggressiveness, try one of the cultivars. Add ribbon grass to your pond, planted in a pot and placed 10–30 cm below the water's surface, to contrast with the leaves of floating water lilies.

Planting

Seeding: not recommended
Transplanting: spring or fall
Spacing: 90 cm

Growing

Ribbon grass grows well in **full sun**
or **partial shade**. The soil should be
of **average fertility** and **moist to
wet;** this grass can grow in water up
to 30 cm deep. It can be invasive and
difficult to remove once established,
though less so in dry locations, so
consider restricting it to a large container to control its spread.

Divide ribbon grass as needed in
spring or early summer. Cut the
plant back to 10 cm tall when it
turns brown in fall.

P. arundinacea 'Feesey's Form'

Tips

This vigorous grass is a great
groundcover. Use the variegated varieties in beds and borders, either
mass planted or as specimens. Use
some form of root barrier, such as
thin metal sheeting or a bottomless
container to at least 30 cm deep, to
prevent the grass from spreading out
of bounds. The flower spikes look
good in fresh and dried arrangements.

Recommended

P. arundinacea is a clump-forming
perennial grass that has dark green
to blue-green foliage and spreads
quickly by rhizomes. **'Feesey's Form'**
has pink-tinged, light green foliage
with wide, white stripes and pale
pink flowers. **Var. *picta*** ('Picta') has
long, narrow, arching, green leaves
with white stripes. The pale pink to
white flowers are held above the foliage in early summer.

Problems & Pests

Ribbon grass is subject to typical
grass family diseases, including leaf
spot, ergot, rust, smut and brown
patch.

P. arundinacea

Rodgersia

Rodger's Flower, Rodger's Plant
Rodgersia

Height: 60 cm–1.8 m • **Spread:** 90 cm–1.2 m • **Flower colour:** white, pink •
Blooms: late spring, summer • **Hardiness:** zones 3–8

Rodgersias are large, bold plants that can fill a big space next to a water feature. Like many big-leaved plants, both these species need consistent moisture to look their best through summer. The new spring foliage of all *Rodgersia* species is often bronze-tinged.

Planting

Seeding: start seeds indoors in late winter or early spring
Transplanting: spring
Spacing: 90 cm

Growing

Rodgersias grow well in **partial** or **light shade**. Excessive exposure causes leaf scorch, so a location that provides protection from winds and the hot afternoon sun is best. The soil should be **fertile, humus rich** and **moist**. These plants don't like to dry out, but they tolerate a drier soil in a shaded location. Divide in spring when necessary.

Tips

These bold plants give a dramatic and exotic appearance to the garden. They thrive in light shade near a pond or other water feature. Rodgersias are also useful as specimen plants or at the back of a bed or border.

R. aesculifolia (both photos)

R. pinnata with *Hakonechloa macra* 'Aureola'

Recommended

R. aesculifolia (fingerleaf Rodgersia) forms a clump of horse chestnut–like leaves 90 cm–1.2 m tall and wide. It produces clusters of tiny, white or pink flowers in mid- or late summer on tall stalks that grow to 1.8 m.

R. pinnata (featherleaf Rodgersia) has large, rough, dark green foliage and white to light pink flowers that bloom in early to mid-summer. Plants grow 60 cm–1.2 m tall and 90 cm–1.2 m wide. **'Chocolate Wing'** is slightly smaller than the species, producing upright stems of chocolate bronze-coloured foliage that turns dark green over the season. It has deep pink to red flowers. (Zones 4–8)

R. aesculifolia

R. podophylla

R. pinnata

R. podophylla (bronzeleaf Rodgersia) has leaves similar to those of fingerleaf Rodgersia, but it differs in that the foliage takes on a bronzy purple cast in mid-summer and fall. This species bears creamy white flowers from late spring to mid-summer and grows 90 cm–1.2 m tall and wide. (Zones 4–8)

Problems & Pests
Rarely, slugs or snails may attack young foliage.

The flower plumes can be cut for use in fresh and dried arrangements.

Rosinweed

Compass Plant, Cup Plant
Silphium

Height: 1.8–3 m • **Spread:** 90 cm–1.5 m • **Flower colour:** yellow •
Blooms: mid- or late summer to fall • **Hardiness:** zones 5–8

Rosinweeds are tall, fast-growing plants that are fairly easy to grow and maintain. The summer flowers attract bees and butterflies, and the seed heads provide food for the local birds. Choose your planting site with care, because rosinweeds can be difficult to remove from an area. The taproot is very long, and new plants will sprout from any sizable piece of the root left in the ground. These plants are worth trying in zone 3, as long as you can provide water and perhaps mulch to protect the crown in winter.

Planting

Seeding: start seeds in late fall or early spring in a cold frame
Transplanting: spring
Spacing: 60 cm

Growing

Rosinweeds grow well in **full sun** or **partial shade**. The soil should be of **average fertility, moist** and **neutral to alkaline**. These plants are tolerant of heavy clay soils. If the soil is too fertile or the site is too shady, rosinweeds may become lanky and floppy. Divide in spring. The rootstocks are massive, so dividing mature plants may be a two-person operation.

Tips

These tall plants can be used in a meadow or prairie garden, or at the edge of a woodland garden. Rosinweeds make interesting additions to bog gardens and waterside plantings.

Recommended

S. laciniatum (compass plant, pilot plant, polar plant) forms a thick clump of tall stems. It grows up to 3 m tall and spreads about 90 cm. The foliage is oriented with the flat side of the leaves facing east or west.

S. perfoliatum (both photos)

Clusters of east-facing, yellow, daisy-like flowers appear in late summer and fall. Be sure to plant this perennial so the flowers face you when the plant is in bloom. This species easily self-seeds.

S. perfoliatum (cup plant) forms a clump of stems about 1.8–2.4 m tall and about 1.2–1.5 m in spread. It bears clusters of yellow, daisy-like flowers from mid-summer to fall. The leaves form a cup around the stem; watch for birds and insects taking a drink.

Problems & Pests

Occasional problems with downy mildew, rust or fungal leaf spot can occur.

Sedge

Carex

Height: 15–90 cm • **Spread:** 30–90 cm • **Flower colour:** green, brown •
Blooms: late spring, summer • **Hardiness:** zones 2–8

There is a little poem that helps you identify a grass-like plant you encounter. It goes "Sedges have edges, Rushes are round, and Grasses are hollow right up from the ground." Sedges have three-sided stems that can be felt if you roll the stems in your fingers. The *Carex* genus belongs to the same plant family as *Cyperus papyrus*, the plant the Egyptians used to make paper, and *Eleocharis dulcis*, the very tasty Chinese water chestnut. Many of these plants prosper with normal moisture circumstances. The sedge family is very large, with around 80 genera and thousands of species.

Planting

Seeding: start seeds in a cold frame in fall, except leatherleaf sedge, which can be started in spring
Transplanting: spring or fall
Spacing: 30–60 cm

Growing

Sedges grow well in **full sun** or **partial shade**. The soil should be **fertile, moist** and **well drained,** though Bowles golden and broad-leaved sedges grow well in wet soil. Old growth can be cut back or thinned by hand in spring to make way for the new growth. Divide overgrown clumps in early summer.

Tips

Sedges make excellent additions to beds, borders and plantings in damp areas of the garden. They are frequently used around the borders of a water feature or directly in the water if they are water-loving species.

C. buchananii (both photos)

Recommended

C. buchananii (leatherleaf sedge) is a dense, tuft-forming evergreen sedge with arching orange-brown, grass-like foliage. It grows 60–75 cm tall and spreads up to 90 cm, bearing spikes of small brown flowers in mid- to late summer. (Zones 6–8)

C. elata **'Aurea'** ('Bowles Golden') forms a clump of arching, grass-like yellow leaves with green margins. It grows 60–90 cm tall and spreads about 46 cm. Spikes of tiny, brown or green flowers appear in early summer. (Zones 5–8)

C. flacca (blue sedge; *C. glauca*) is a durable, slowly creeping, variable sedge with narrow, blue-grey foliage. Plants do well in a range of climate conditions and are evergreen in the warmer zones. The wispy flowers bloom in early to mid-summer and are held above the foliage. The foliage grows 15–30 cm tall and 30–46 cm wide. (Zones 4–8)

C. flagellifera forms tall, dense tufts of graceful, arching, narrow, bronze red to reddish brown foliage that grows 41–51 cm tall, spreads 46–60 cm and is evergreen in the warmer zones. The tan to reddish brown flowers reach 51–90 cm. (Zones 4–8)

C. grayi (Gray's sedge, morning star sedge) has shiny, wide, medium to dark green, semi-evergreen foliage, interesting spherical, spiky light green flowers beginning in late spring and brown seed heads. Plants grow 30–90 cm tall and 30–60 cm wide.

C. morrowii (creeping Japanese sedge) **'Ice Dance'** is a graceful, vigorous selection that has wide, shiny, dark, semi-evergreen foliage with creamy white margins. Plants grow 25–36 cm tall and 30–46 cm wide. (Zones 4–8)

C. oshimensis (*hachijoensis*) **'Evergold'** (variegated Japanese sedge) forms a low mound of evergreen, grass-like, dark green-and-yellow-striped foliage. It grows about 46 cm tall and spreads about 30 cm. It bears spikes of small brown flowers in late spring and early summer. (Zones 4–8)

C. oshimensis 'Evergold'

C. elata 'Aurea'

***C. siderosticha* 'Variegata'** (variegated broad-leaved sedge) forms a clump of pale green, wide, strap-shaped leaves with creamy margins and pink-flushed bases. It grows about 30 cm tall and spreads about 46 cm. It produces spikes of light brown flowers in late spring. (Zones 4–8)

Sedges are among the most attractive grass-like plants, often used for the moist, pondside garden. In the wild, their dense, tufted clumps can mislead hikers into believing the ground is more solid than it actually is.

Problems & Pests
Rare problems with aphids, fungal leaf spot, rust and smut can occur.

Self-Heal

Prunella

Height: 10–30 cm • **Spread:** 15–46 cm • **Flower colour:** purple, white, pink • **Blooms:** late spring, summer, fall • **Hardiness:** zones 3–8

Self-heal is a hearty member of the mint family and boasts a long blooming period, wonderful flower colour and the ability to attract beneficial insects. This plant can be short-lived, sometimes acting more like a biennial than a perennial. It self-seeds freely and can be invasive, ensuring you will have this little gem for a long time.

Planting

Seeding: grows easily from seed; sow directly into warm soil in spring, or sow indoors in spring or fall; cold stratification aids germination; plants often bloom the first year from seed

Transplanting: spring or fall

Spacing: 46 cm

Growing

Self-heal grows best in **full sun to partial shade** in **average to moist, well-drained** soil but will do well in almost any soil. Self-heal does not thrive in dry conditions and appreciates regular watering. Deadhead to prevent self-seeding. For large areas, use a bagging lawnmower. Propagate this plant by division in spring or fall, by stem cuttings and by layering the stem.

Tips

Self-heal's spreading, vigorous nature makes it a good groundcover. It is works well in containers where the spread is easily controlled, or in a wildflower garden where spreading is not as much of an issue. It can be used as edging in perennial or mixed beds and borders, but be aware that it can engulf less vigorous plants. Self-heal will invade lawns, and it can be used as a lawn substitute. It will adapt to the height of cut and will bloom just under that height. The flowers attract bees and other beneficial insects.

Recommended

P. grandiflora (large self-heal) is a vigorous, spreading, low-growing perennial that roots where stem nodes touch the ground. It bears dark green foliage and grows 10–30 cm tall and 30–46 cm wide but can spread 90 cm or more. Dense, upright, long-lasting clusters of two-lipped, light and darker purple flowers bloom in late spring to late summer. **'Blue Loveliness'** bears light purple flowers. **'Pink Loveliness'** bears flowers in shades of pink. **'White Loveliness'** has white flowers. (Zones 4–8)

P. vulgaris (common self-heal) is an erect to creeping plant whose stems root at the nodes. It bears aromatic, lance-shaped foliage. This plant grows 10–30 cm tall and spreads 15–30 cm or more. It produces dense clusters of whitish purple to pinkish purple flowers in late spring to early fall.

Problems & Pests

Self-heal rarely suffers from any problems but might experience powdery mildew, leaf spot, slugs and snails.

P. grandiflora 'Pink Loveliness'

Sheep's Bit

Shepherd's Scabious
Jasione

Height: 20–30 cm • **Spread:** 20–30 cm • **Flower colour:** blue to purple-blue • **Blooms:** early to late summer • **Hardiness:** zones 4–8

The flowers of sheep's bit resemble those of some of my favourite plants in the Teasel family *(Dipsacaceae)*, such as knautia, but sheep's bit is not related to that family. It is part of the bellflower family and is an easy-to-grow, low-maintenance perennial, as are many of the bellflowers. It is a great plant for those new to perennial gardening because it grows easily from seed, and dividing the plant is simple. Sheep's bit may act somewhat like a biennial in colder zones.

Planting

Seeding: sow seeds in containers in a cold frame in spring or fall; plants may bloom in the first year
Transplanting: spring or fall
Spacing: 20–30 cm

Growing

Sheep's bit grows best in **full sun** in **moderately fertile, average to moist, well-drained, neutral to acidic, loamy to sandy** soil. Plants are tolerant of partial shade and do better with winter protection in the colder hardiness zones. Deadhead to encourage even more blooms. Plants can be propagated by division in spring and by taking stem cuttings.

Tips

Sheep's bit looks great when mass planted or as edging in beds and borders. It is also attractive in rock, scree and alpine gardens, meadow and wildflower gardens and dotted in cottage gardens. The flowers are good for cutting. Try sheep's bit in containers, especially in areas where amending the existing soil to provide the best growing conditions is near impossible.

Recommended

J. laevis forms dense, compact tufts of narrow, grey-green basal foliage. Nearly spherical, pincushion-like clusters of light purple-blue to blue flowers are held above the foliage on upright to nodding stems. Flowering begins in early summer, and plants bloom continuously until late summer. Plants usually grow 20–30 cm tall and wide but may grow a little

J. laevis (both photos)

taller in ideal conditions. **'Blue Light'** bears bright to deep blue flowers.

Problems & Pests

Snails and slugs may eat the young foliage.

Shooting Star

Dodecatheon

Height: 15–30 cm **Spread:** 15–25 cm • **Flower colour:** pink, white •
Blooms: spring • **Hardiness:** zones 4–8

Shooting star is one of the earliest sun-worshipping perennials to appear in spring. This petite plant puts up erect stems topped by tiny flowers whose petals are swept back, giving the flowers the appearance of shooting stars, hence the common name. It is worth trying in zone 3.

Planting
Seeding: not recommended
Transplanting: spring or fall
Spacing: 15–30 cm

Growing
Shooting star prefers **partial shade** with shade in the afternoon but tolerates full sun. It grows best in soil that is **moist, humus rich** and **well drained**. However, being a prairie native, it tolerates many soil conditions.

This plant goes dormant in summer. Provide plenty of water when it is actively growing, but hold back after flowering because it enters dormancy.

Division is possible in late winter or early spring, but this plant does not like being disturbed. Try cutting out chunks of soil with the roots and transplanting the whole chunk.

D. meadia (both photos)

D. meadia (all photos)

Tips

Shooting star works well in a woodland garden, wild garden or rock garden. Keep shooting star separate from taller plants or it will be crowded out. It makes a fine specimen.

In the Prairies, this plant has finished flowering by the time larger grasses and herbaceous plants take over. In the garden bed, make sure other spring-flowering perennials do not overtake it.

Recommended

D. meadia forms clumps of foliage with the flowers held above the foliage. It produces pale to deep pink

flowers in spring. **'Alba'** has white blooms.

Problems & Pests
Occasional problems with rust, slugs and snails can occur.

If you attempt to grow this plant from seed, as some adventurous gardeners do, it will need cold stratification to break the seeds' dormancy. Sow ripened seed in the garden or in flats in a cold frame in late summer or fall.

Skullcap

Scullcap
Scutellaria

Height: 10–90 cm • **Spread:** 20–46 cm • **Flower colour:** blue-violet, blue, purple, white, rose, pink, pale yellow • **Blooms:** summer, early fall • **Hardiness:** zones 4–8

Skullcap is another wonderful, colourful, multipurpose plant that is under-used in Canadian landscapes and gardens. Plants in less than ideal soil may act like biennials, so allow some self-seeding to occur. *S. alpina* can flower in the first year from seed.

Planting

Seeding: sow fresh seeds directly in fall, in containers in a cold frame in fall or early spring, or indoors in late winter; cold stratification is recommended; most species need about one month of cold, but *S. incana* needs at least 3 months
Transplanting: spring
Spacing: 25–46 cm

Growing

Skullcap grows best in **full sun to light shade** in **well-drained, light, sandy to gravelly, neutral to slightly alkaline** soil. Let the soil dry a little bit between watering. Provide winter protection in colder zones. Propagate by division in spring or fall, or by taking basal or stem cuttings in mid-spring to early summer. Propagate *S. incana* only by stem cuttings or seed because it does not transplant well when mature.

Tips

Skullcap looks good as an accent or when planted en masse and is a natural in a herb garden. It is also a good container plant. Use *S. alpina* and *S. pontica* in rock and alpine gardens and for edging beds and borders. Use *S. incana* in beds, borders, cottage gardens, wildflower, meadow and prairie plantings, and at the edges of woodlands.

Recommended

S. alpina is a slow-spreading, tuft-forming, hairy plant that roots where the stem nodes touch the ground. It has dark grey-green foliage and grows 15–25 cm tall and 20–30 cm wide. Dense clusters of two-lipped, purple flowers with white to creamy white lower lips at the tips of the flower stems are borne in summer. **'Arcobaleno'** bears bicoloured flowers in shades of blue-violet, pale blue, white, rose, pink and pale yellow.

S. incana (downy skullcap, hoary skullcap) grows 60–90 cm tall and 30–46 cm wide. It has long, ovate, crinkly, medium to dark green leaves with toothed margins. Most of the plant, other than the upper leaf surfaces, is covered in fine hairs. Dense clusters of large, blue to purple-blue flowers bloom from mid- to late summer.

S. pontica (Turkish skullcap) is a low-growing, mat-forming plant with rosettes of oval, fuzzy, grey-green foliage. It bears violet to rosy purple flowers in mid- to late summer. Plants grow 10–20 cm tall and 20–30 cm wide. (Zones 5–8)

Problems & Pests

Skullcap is relatively pest free but may experience powdery mildew and aphids.

S. alpina

Stokes' Aster

Stokesia

Height: 30–60 cm • **Spread:** 30–46 cm • **Flower colour:** purple, blue, white, pink, pale yellow • **Blooms:** mid-summer to early fall • **Hardiness:** zones 4–8

Stokes' aster blooms arise on somewhat lax stems. This easygoing nature, the long-lasting late-season bloom, good pest resistance, evergreen foliage in the warmer zones and the variety of flower colours make Stokes' aster a winner in your garden. Stokes' aster with blue or white blooms is commonly found at nurseries. It is also worth your while to seek out the yellow-flowered 'Mary Gregory.' The blooms attract butterflies.

Planting

Seeding: start seeds indoors or outdoors in early fall into warm soil
Transplanting: spring
Spacing: 30 cm

Growing

Stokes' aster grows best in **full sun**. The soil should be **average to fertile, light, moist** and **well drained**. This plant dislikes waterlogged and poorly drained soils, particularly in winter when root rot can quickly develop. Those who garden in clay might consider planting this species in raised beds filled with carefully amended soil. Provide a good winter mulch to protect the roots from the cycles of freezing and thawing, particularly in zones 4 and 5. This plant self-sows but not aggressively. Divide in spring.

S. laevis

S. laevis 'Colorwheel'

S. laevis

Deadheading extends the bloom, which can then last up to 12 weeks.

Tips

Stokes' aster can be grouped in borders and adds some welcome blue to the garden late in the season when yellows, oranges and golds seem to dominate. It makes an excellent cut flower, staying open for a week or more. It is also good as a container plant.

Recommended

S. laevis forms a basal rosette of bright green, narrow foliage. The midvein of each leaf is a distinctive pale green. The plant bears purple, blue or sometimes white or pink flowers from mid-summer to early fall. **'Alba'** has white flowers. **'Blue Danube'** bears large purple-blue flowers. **'Colorwheel'** produces flowers that open white and mature to deep purple, showing every colour in

S. laevis 'Rosea'

S. laevis 'Blue Danube'

between. Flowers in shades of pink, purple and white appear, sometimes all at once on the plant, creating a virtual tapestry. **'Klaus Jelitto'** bears large, light blue flowers. **'Mary Gregory'** is a compact cultivar with light yellow flowers. **'Omega Skyrocket'** is a strong-growing plant, 90 cm–1.2 m tall, that bears large purple flowers. **'Peachie's Pick'** is a dense-growing, compact plant that bears blue flowers. **'Purple Parasols'** has flowers that open blue and mature to purple. **'Rosea'** produces pink flowers.

Problems & Pests

Problems are rare, though leaf spot and caterpillars can cause trouble. Avoid very wet and heavy soils to prevent root rot.

Stonecress

Persian Stonecress
Aethionema

Height: 5–25 cm • **Spread:** 15–30 cm • **Flower colour:** light pink, white •
Blooms: early spring to mid-summer • **Hardiness:** zones 4–8

Stonecresses are tough, short-lived, evergreen to semi-evergreen perennials
or subshrubs native to the mountains of Europe through the Middle East to
Western Asia. They self-seed in favourable conditions, so deadhead the plants
after blooming unless you want the plants to spread. Plant them with whitlow's grass (*Draba*) for early-spring colour.

Planting

Seeding: start seeds in containers in a cold frame in spring, or direct sow when the soil temperature reaches 4° C
Transplanting: spring
Spacing: 15–25 cm

Growing

Stonecresses grow well in **full sun** in very **well-drained, neutral to alkaline** soil of **average fertility**. They can handle poor soils, dry conditions and moderate drought but will appreciate an occasional watering. Cut plants back lightly after flowering has finished to maintain a bushy habit. Mulch plants deeply in the colder regions if snowcover is unreliable. Take softwood cuttings in late spring or summer to propagate more plants.

Tips

Stonecresses are excellent plants for alpine gardens and troughs, rock gardens and stone walls. They do well in containers, and the taller species make great edging plants.

Recommended

A. cordifolium is a low-growing, mound-forming plant with red-tinged, branching stems and narrow, blue-green to blue-grey foliage. It grows 15–25 cm tall and 25–30 cm wide and bears clusters of small, light rose-pink flowers from late spring to mid-summer.

A. grandiflorum forms low-growing mounds of soft, fleshy, narrow, blue to blue-green foliage. Clusters of small, fragrant, light pink flowers bloom from mid-spring to early

A.oppositifolium (both photos)

summer. It grows 15–20 cm tall and 25–30 cm wide.

A. oppositifolium (*Eunomia oppositifolia*) is a dense, low-growing, mounded plant with fleshy, rounded, grey-green foliage. It grows 5–8 cm tall and 15–20 cm wide. From early to mid-spring, plants produce a plethora of fragrant, white to light pink flowers that arise from attractive purple flower buds.

Problems & Pests

Stonecresses have occasional problems with aphids and spider mites.

Toad Lily

Tricyrtis

Height: 20–90 cm • **Spread:** 20–60 cm • **Flower colour:** white, pink or purple, with red or purple spots • **Blooms:** late summer and fall • **Hardiness:** zones 4–8

The tiny, interesting flowers of toad lilies need to be viewed closely to appreciate their beauty. Place the plants where you can approach them for a good look at the flowers, such as at the edge of a woodland walk, at the base of a low window or by your doorway. The small flowers bloom in fall and run the length of the arching, leafy stems. The species and cultivars hybridize easily, producing original progeny. Planting toad lilies in masses will provide the best show.

Planting

Seeding: sow ripe seeds indoors or in a cold frame in early spring
Transplanting: spring
Spacing: 30–60 cm

Growing

Toad lilies grow well in **partial to light shade** in a **sheltered location**. They tolerate full shade, but flowering may be delayed. The soil should be **fertile, humus rich, moist** and **well drained**. The foliage may suffer tip burn if the plants are under stress or if they get too hot, but it won't harm the plants. A deep, loose layer of mulch should be applied in fall for winter protection, with or without reliable snow coverage. Divide the plants in early spring when they are dormant.

T. hirta 'Hatatogisa'

T. hirta 'Moonlight'

T. hirta

T. hirta 'Togen'

Tips

Toad lily is well suited to woodland gardens and shaded borders, rock gardens, patios and ponds. It can also be grown in containers when the perfect location just simply cannot be found.

Recommended

T. **'Empress'** forms a large, upright clump that grows about 75 cm tall and spreads 46–60 cm. It bears large white flowers with dark reddish purple spots.

T. formosana (formosa toad lily) is an upright spreading plant. It grows 60–90 cm tall and spreads 46–60 cm. The early-fall flowers are white, sometimes tinged pink or purple, with red or purple spots. **'Amethystina'** bears pale purple flowers with

T. hirta 'Miyazaki'

white throats and red spots. **'Samurai'** is a compact plant with yellow-margined leaves and light purple flowers with darker purple spots.

T. hirta (Japanese toad lily) forms a clump of light green leaves. In late summer and fall, it bears purple-spotted white flowers. It grows 60–90 cm tall and spreads 30–60 cm. **Var. *alba*** bears green-tinged white flowers with pinkish anthers. **'Hatatogisa'** bears soft blue flowers spotted in purple with white centres. **'Miyazaki'** has white flowers that are spotted with light purple. The leaves have lighter margins. **'Moonlight'** is a golden-leaved sport of 'Variegata.' It grows about 51 cm tall, with white flowers flecked occasionally with lavender.

'Shirohotogisu' produces pure white flowers, and **'Togen'** has purple flowers with white centres. **'Variegata'** has narrow creamy edges on the green foliage.

***T.* 'Moonlight Treasure'** grows 20–30 cm tall and wide and has dense, olive green, faintly spotted, thick, leathery leaves. The upward-facing, buttery yellow flowers are large in relation to the plant.

***T.* 'Sinonome'** forms a dense clump up to 90 cm tall and 46–60 cm wide. Flowers are purple-and-white spotted.

Problems & Pests
Slugs and snails may cause problems in spring.

Trillium

Wake Robin
Trillium

Height: 30–51 cm · **Spread:** 20–30 cm or more · **Flower colour:** white, yellow, pink, red, purple · **Blooms:** spring · **Hardiness:** zones 4–7

Trilliums are a welcome sight in early spring, but their beauty might not last long. Trilliums go dormant if the soil dries out, so don't dig them out if they suddenly seem to disappear. These beautiful little woodland plants require very little care, but provide a little extra TLC for the first couple of years. You will be happy you did. Never dig plants from the wild, and only buy from reputable nurseries that propagate from seed. Many species of trillium are threatened or endangered because of people collecting these and other rare plants.

Planting

Seeding: not recommended; may take two or more years before signs of growth appear and another five or more years before plants reach flowering size; seeds should not be allowed to dry out before planting; ripe seeds may be started in a shaded cold frame in late summer
Transplanting: fall or spring
Spacing: 30 cm

Growing

Place trilliums in **full** or **partial shade**. The soil should be **humus rich, moist, well drained** and **neutral to acidic**. Rhizomes should be planted about 10 cm deep. Add organic matter, such as compost or well-aged manure, to the soil when planting, and include a mulch of shredded leaves to encourage rapid growth. Division is not necessary. Apply compost every year.

T. grandiflorum with lily-of-the-valley mixed in around the base

T. grandiflorum

T. sessile

T. undulatum

Tips

These plants are ideal for natural woodland gardens and for plantings under spring-flowering trees and shrubs.

Trilliums are best left alone once planted. New transplants may take a year or two to adjust and start flowering. The amount of moisture they receive after flowering greatly influences how quickly the plants establish. Plentiful moisture in summer prevents the plants from going dormant after flowering. Instead, they send up side shoots that increase the size of the clump and the number of flowers the following spring.

Recommended

T. erectum (purple trillium, red trillium) has deep wine-red flowers. It grows up to 51 cm tall and spreads up to 30 cm. When the flower is spent around June, it produces a red berry.

T. grandiflorum (great white trillium, snow trillium) has large white flowers that turn pink as they mature. It grows 41 cm tall and spreads 30 cm or more. When the flower dies back, a black fruit follows. **'Flore Pleno'** has double flowers but is slower growing.

T. luteum (yellow whippoorwill, golden goblet trillium) forms a clump of dark green leaves spotted with lighter green. It bears fragrant bright yellow flowers in spring. Plants grow about 41 cm tall and spread about 30 cm.

T. erectum

T. recurvatum (purple wake robin, purple trillium, bloody butcher) has dark purple or occasionally yellow or white flowers. It grows up to 41 cm tall and spreads 30 cm.

T. sessile (toadshade trillium) forms a clump of dark green leaves mottled with silver, bronze, maroon and lighter green. It bears dark purple-red flowers in late spring. Plants grow about 30 cm tall and spread about 20 cm.

T. undulatum (painted trillium) has pure white, upward-facing flowers with a prominent triangle of red-purple in the throat that bleeds up along the veins of the petals to the tips. The fruit is a distinctive, smooth oval berry with a pointed tip. Plants grow 30–51 cm tall and 20–25 cm wide. They are difficult to grow and are rare.

Problems & Pests
Trilliums have few pest problems, but snails and slugs favour the young foliage.

Tuber Oat Grass

Arrhenatherum

Height: 15–60 cm • **Spread:** 30–41 cm • **Flower colour:** purple-tinged, silvery green • **Blooms:** mid-summer to early fall • **Hardiness:** zones 3–8

Tuber oat grass is a tough, relatively low-maintenance ornamental grass that begins growing early in spring, tolerates exposure to the elements and does well in urban settings. It is a cool-season grass that grows heartily in spring and fall and can go dormant in summer, much like an unwatered Kentucky bluegrass lawn. If your plant goes dormant, it will return in late summer to fall as the air temperature cools. Tuber oat grass is a spreading grass and can be invasive. It can easily escape from cultivated landscapes.

Planting

Seeding: sow seeds directly in fall, or in containers in a cold frame in fall or early spring
Transplanting: spring
Spacing: 30–46 cm

Growing

Tuber oat grass grows best in **full sun to partial shade** in **moist, well-drained, fertile, neutral to slightly acidic** soil. Plants tolerate a range of soils as long as they are well drained. Allow the soil to dry out a little between watering. Mature clumps are somewhat drought tolerant. Plants will suffer in heavy, wet soils. Cut back shabby-looking plants in summer to promote new growth. Cut large plantings back with a bagging lawnmower set at the highest cutting height. Divide plants in spring every three to four years to maintain vigour and to propagate more plants.

Tips

Tuber oat grass is attractive when mass planted in beds and borders and can be used as edging or as a groundcover. It looks great in a rock garden. It also makes a nice accent plant in containers.

Recommended

A. elatius var. *bulbosum* 'Variega-tum' forms upright, spreading tufts of grey-green to blue-green foliage variously striped and edged with white. Insignificant, narrow, spiky clusters of small, purple-tinged, green flowers bloom on slightly arching stems from mid-summer to

A. elatius var. *bulbosum* 'Variegatum' with other grasses

early fall. Plants may not produce flowers if the summer is too hot.

Problems & Pests

Tuber oat grass may experience fungal diseases, such as rust, in hot, humid climates.

A. elatius var. *bulbosum* 'Variegatum'

Tulip

Tulipa

Height: 10–56 cm • **Spread:** 5–20 cm • **Flower colour:** all shades except blue • **Blooms:** spring • **Hardiness:** zones 3–8

Tulips have an excellent range of flower colours and shapes and are great plants to use in a perennial design. They provide early colour while your later-flowering perennials are developing their foliage. Then as the later-flowering perennials fill out and bloom, they cover the fading tulips. Some tulips are more perennial than others, including species tulips such as *T. tarda* and *T. sylvestris* and their relatives. Selections of Fosteriana, Kaufmanniana, Greigii and Darwin Hybrid tulips are all longer lived.

Planting

Seeding: not recommended
Transplanting: fall
Spacing: 5–20 cm

Growing

Tulips grow best in **full sun**. The flowers tend to bend toward the light in light or partial shade situations. The soil should be **fertile** and **well drained**. Plant bulbs in fall, 10–15 cm deep. Bulbs that have been cold treated may be available for purchase in spring and can be planted at that time. Although many tulips come back and bloom each spring, many hybrids perform best if planted new each year. The species, older cultivars and the types mentioned above are the best choices for naturalizing.

Tips

Tulips provide the best display when mass planted or planted in groups in flowerbeds and borders. They can also be grown in containers and can be forced to bloom early in pots indoors. Some tulips can be naturalized in cottage, meadow and wildflower gardens.

Recommended

T. batalinii has grey-green leaves with wavy reddish edges. It grows 20–30 cm tall. Small bowl-shaped flowers in various shades of yellow, apricot and red bloom late April to early May.

T. **'Blushing Apeldorn'** is a Darwin Hybrid tulip. It has large, cup-shaped, lemon-yellow flowers with persimmon-orange edges and interior feathering. It grows 56 cm tall and blooms from mid-April to May.

T. **'Heart's Delight'** is a Kaufmanniana tulip, also known as a waterlily tulip because the petals open flat, giving the appearance of a waterlily. It grows 25 cm high. The early-April flowers are rich carmine red on the outside of the petals, white fading over time to pale rose on the interior and have a yellow centre.

T. **'Little Beauty'** has pinkish red, star-shaped, early-May flowers with dark blue centres lined in pale pink. It grows 10–15 cm tall.

T. **'Little Princess'** has dark orange, star-shaped, early-May flowers with dark blue centres lined in yellow. It grows 10–15 cm tall.

T. 'Heart's Delight'

T. **'Purissima'** ('White Emperor') is a Fosteriana tulip with pure white, fragrant flowers with immense petals and is fantastic planted in large drifts or used in bridal bouquets. It grows 46 cm tall and blooms in April or May.

T. **sylvestris** spreads by stolons, has light green leaves and grows in sun or shade. It bears fragrant, star-shaped, yellow or cream-coloured flowers in April. It grows about 41 cm tall.

T. **tarda** is very easy to grow and also spreads by stolons. It has narrow, glossy green leaves and star-shaped white, yellow-centred flowers, sometimes five to a stem. The plants flower abundantly from April through May. They grow only 10–15 cm tall.

T. sylvestris

T. batalinii

T. 'Toronto' is a multi-flowering (several blooms per stem) Greigii tulip with purple-and-green-striped leaves. The bowl-shaped March or April flowers are pinkish red with orange-red insides and yellow centres. It grows 20–30 cm tall.

Problems & Pests

Bulb rot can occur in poorly drained soil, and slugs, aphids and nematodes may attack plants.

During the "tulipomania" of the 1630s, the bulbs were worth many times their weight in gold, and many tulip speculators lost massive fortunes when the mania ended.

Tulip flowers come in several forms—they can be single or double and may be bowl-shaped, lily-shaped, cup-shaped, goblet-shaped or star-shaped. They can also have fringed or long, narrow petals.

Turtlehead

Chelone

Height: 46 cm–1.2 m • **Spread:** 46–90 cm • **Flower colour:** pink, purple, white •
Blooms: late summer and fall • **Hardiness:** zones 3–8

Turtleheads are North American natives that are exceedingly easy to grow.
These worthy but underused perennials have glossy, dark green foliage and
lovely clusters of softly coloured, rounded flowers that can bloom well into
September, when fresh new bloom is needed most. White turtlehead is a
moisture-lover, and you may spot it growing in wild, boggy areas from
Manitoba eastward.

Planting
Seeding: sow seeds indoors in late winter, or outdoors in late spring into warm soil
Transplanting: spring
Spacing: 46–60 cm

Growing
Turtleheads grow best in **partial shade** or **full sun.** The soil should be **fertile, humus rich, moist** and **well drained,** but these plants tolerate clay soil and boggy conditions. Plants may become weak and floppy in too shaded a spot, so pinch their tips in spring to encourage bushy growth. Divide plants in spring or fall. They can be propagated from stem cuttings taken in early summer.

C. obliqua (both photos)

Tips
Turtleheads can be used in a pond-side or streamside planting. They also do well in a bog garden or in a moist part of the garden where plants requiring better drainage won't grow.

Recommended
C. lyonii (pink turtlehead) is an erect plant with square stems. This species is almost identical to *C. obliqua*, except that the flowers are a slightly darker pink, blooming from late summer to fall, and the plant is a little larger. *C. lyonii* is more commonly grown and is possibly the more adaptable species. Plants grow 60 cm–1.2 m tall and 60–90 cm wide.

C. obliqua (rose turtlehead) is an upright plant that grows 46–90 cm tall and 46–60 cm wide, forming a dense mound of foliage. From late summer to fall, the plants bear pink or purple flowers. **'Alba'** (white turtlehead; *C. glabra*) bears white flowers slightly earlier than the species.

Problems & Pests
Rare problems with powdery mildew, rust and fungal leaf spot can occur.

Umbrella Plant

Darmera

Height: 30 cm–1.8 m • **Spread:** 90 cm–1.5 m • **Flower colour:** white to pink • **Blooms:** mid- to late spring • **Hardiness:** zones 5–8

Picture the little saxifrage you have in your garden. Now picture it on steroids and you have the umbrella plant. It has interesting flowers and stunning foliage that turns wonderful shades of red in fall. Umbrella plant will perform well in your garden if you can keep the soil moisture at a proper level. The plant can grow in shallow water, so don't worry about overwatering. Make sure you give this eye-catching plant room to grow and spread. It is definitely worth trying in zones 3 and 4.

Planting

Seeding: sow seeds directly in fall or in containers in a cold frame in fall or early spring; ensure the soil remains moist; germination may be slow
Transplanting: in spring, after the risk of frost has passed
Spacing: 60 cm–1.2 m for species; 40–51 cm for 'Nana'

Growing

Umbrella plant grows well in **full sun to partial shade** in **moist to wet, moderate to fertile** soil. Keep the soil consistently moist, especially if the plant is grown in drier areas or in full sun. Leaf margins may scorch if the plant gets too dry. The new growth in spring may be damaged by frost. Division is rarely needed but can be done in spring to propagate more plants.

Tips

Umbrella plant makes a wonderful specimen and is striking when massed. It is a natural for bog gardens and near ponds and streams, where the rhizomes and water-collecting leaves help stabilize the muddy soils. It can also be effective in beds and borders and can be allowed to ramble through a woodland. The flowers attract bees and butterflies and can be cut for fresh and dried arrangements.

Recommended

D. peltata (*Peltiphyllum peltatum*) forms 60 cm–1.2 m tall and 90 cm–1.5 m wide clumps of large, basal, rounded, deeply lobed, prominently veined, dark green leaves up to 60 cm across. It spreads slowly by

D. peltata (both photos)

thick rhizomes. The leaves turn bronze to red in fall. Rounded clusters of white to pink flowers emerge before the foliage on tall, thick, slightly hairy stems that grow up to 1.8 m. **'Nana'** is a compact version that grows 30 cm tall and 41–60 cm wide, and the flower stems can reach 60 cm. Each leaf can be up to 30 cm wide.

Problems & Pests

Umbrella plant rarely suffers pest problems.

Virginia Bluebells

Cowslip
Mertensia

Height: 30–60 cm • **Spread:** 25–46 cm • **Flower colour:** blue, purple-blue, pink • **Blooms:** mid- and late spring • **Hardiness:** zones 3–8

You might forget you have planted Virginia bluebells until their foliage arises in spring. Pink buds follow the foliage, opening to blue flowers. After the blooms disappear, the foliage begins to yellow and is gone by mid-summer. The disappearing foliage is a normal attribute, not something to worry about. Use a weather-resistant marker, such as a stone with 'Virginia Bluebells' painted on it, to remind yourself where in your garden this plant grows.

Planting

Seeding: start in a cold frame in fall
Transplanting: spring or fall
Spacing: 30–46 cm

Growing

Virginia bluebells grow best in **light to partial shade** with shelter from the hot afternoon sun. Plants perform fairly well in full shade as long as it is not too deep. The soil should be of **average fertility, humus rich, moist** and **well drained**. Plants self-seed in good growing conditions. Virginia bluebells go dormant in mid-summer and can be divided at that time or in spring, just as new growth begins. They can be propagated by root cuttings in late fall to early winter.

Tips

Include Virginia bluebells in a shaded border or moist woodland garden. Make sure to plant them with plants that cover the bare spaces left when Virginia bluebells go dormant by mid-summer. They form large colonies if left to their own devices.

Recommended

M. virginica (*M. pulmonarioides*) forms an upright clump with light green to blue-green leaves. It bears clusters of blue or purple-blue flowers that open from pink buds in late winter to early spring. **'Rosea'** produces rose-pink flowers.

Problems & Pests

Virginia bluebells have infrequent problems when grown in good conditions. Plants may experience slugs, snails, powdery mildew and rust.

M. virginica

Virginia bluebells are quite fragile, so take care when working around the plants.

M. virginica and *Epimedium*

Whitlow's Grass

Draba

Height: 5–10 cm • **Spread:** 8–25 cm • **Flower colour:** yellow • **Blooms:** early to mid-spring • **Hardiness:** zones 2–8

Draba is a large, diverse genus of tough, low-maintenance plants from a variety of climates, many from harsh northern climates, and native *Draba* species are found all over Canada. They are some of the earliest plants to bloom in spring. I grow *D. aizoides,* and the bright yellow blooms are my signal to begin the spring gardening processes. If you have to transplant them, take as deep a rootball as possible.

Planting

Seeding: sow seeds directly or in containers in a cold frame in spring or fall; seeds can also be sown indoors in spring
Transplanting: spring
Spacing: 10–30 cm

Growing

Whitlow's grass grows best in **full sun** in **well-drained, gravelly to sandy** soil but will grow in a range of soils as long as there is excellent drainage. *D. aizoides* prefers neutral to slightly alkaline soil, and *D. rigida* does well in slightly acidic soil. This grass needs regular watering, but allow the soil to dry in between water applications. It is drought tolerant when established. Protect it from winter moisture and from standing water. Division is difficult because of the deep roots. Propagate by separating and planting individual rosettes or by stem cuttings in spring.

Tips

Use this diminutive plant in rock, scree and alpine gardens, raised beds, containers and for miniature theme gardens. It can also be used to edge small beds and borders.

Recommended

D. aizoides is semi-evergreen to evergreen and forms cushion-like mounds of narrow, dark green foliage. Dense clusters of small, bright yellow flowers are held above the foliage and bloom in early to mid-spring. Plants grow 5–10 cm tall and 15–25 cm wide. (Zones 3–8)

D. sibirica

D. rigida forms small cushions of dark green foliage and grows 5–10 cm tall and 10–20 cm wide. Clusters of bright yellow flowers bloom in early to mid-spring. **Var. *bryoides*** is a smaller, tighter plant than the species, with smaller, dense, incurved foliage. It grows 5–8 cm tall and 8–15 cm wide.

Problems & Pests

Whitlow's grass has infrequent occurrences of rust, powdery mildew, aphids and spider mites.

D. aizoides

Wood Fern

Dryopteris

Height: 30 cm–1.2 m • **Spread:** 30 cm–1.2 m • **Flower colour:** grown for foliage; *D. erythrosora* has conspicuous red spore structures • **Hardiness:** zones 3–8

From the lacy silhouette of the fronds to the plants' graceful, arching forms, wood ferns are truly elegant. They are excellent low-maintenance plants that look great when massed in a shady area. They combine well with blooming shade perennials, providing contrast in texture and colour. Once established, these ferns are very tough.

Planting

Seeding: sow spores into warm soil as soon as they are ripe
Transplanting: spring or fall
Spacing: 30–90 cm

Growing

Wood ferns grow well in **light** or **partial shade** but tolerate full sun (with afternoon shade in consistently moist soil) or deep shade. The soil should be **fertile, humus rich** and **moist,** though some species are fairly drought tolerant. Plants can be divided in spring or fall to propagate them or control their spread.

Tips

These large, impressive ferns are useful as specimens or grouped in a shaded area or a woodland garden. They are ideal for an area that stays moist but not wet, and they beautifully complement other shade-loving plants.

D. erythrosora

D. marginalis

D. marginalis

Recommended

D. affinis 'Cristata' (golden-scale male fern) forms a large clump of arching evergreen fronds. The leaflets on the fronds are divided at the tips, giving the plant a fluffy appearance. It grows about 90 cm tall, with an equal spread. (Zones 4–8)

D. x australis (Dixie wood fern) forms an upright clump of semi-evergreen fronds. It grows up to 1.2 m tall and spreads about 60 cm. (Zones 5–8)

D. cristata (crested wood fern) is a shade-loving wetland fern. It forms a small clump of very upright, narrow, airy fronds. Fertile fronds are deciduous, and sterile fronds are

D. erythrosora

D. x australis

evergreen. It grows 30–90 cm tall and spreads about 30 cm.

D. erythrosora (autumn fern) forms upright tufts of shiny, dark green deciduous fronds that contrast with the coppery coloured young fronds. It grows about 60 cm tall and spreads about 46 cm. The large, red spore structures on the undersides of the fronds are conspicuous. Cultivars with frond variations are available. (Zones 5–8)

D. filix-mas (male fern) is similar in appearance to *D. affinis*, but the fronds are lighter in colour, less leathery and are not evergreen. This large fern, with fronds up to 1.2 m long, is one of the best-known wood ferns. (Zones 4–8)

D. goldiana (giant wood fern), one of the largest wood ferns, spreads slowly, producing tufts of deciduous fronds along the rhizome. It grows about 1.2 m tall and spreads 60 cm–1.2 m.

D. marginalis (leatherwood fern, marginal wood fern) forms a large clump of evergreen fronds. It grows 46–90 cm tall and spreads about 60 cm.

Problems & Pests
Occasional problems with leaf gall, fungal leaf spot and rust may arise. Crown rot can occur in poorly drained soils.

Wood Poppy

Celandine Poppy, Flaming Poppy
Stylophorum

Height: 30–46 cm • **Spread:** 25–30 cm • **Flower colour:** yellow • **Blooms:** mid-spring to early summer • **Hardiness:** zones 4–8

Wood poppies are easy-to-grow, long-lived members of the poppy family that have great foliage and bright, cheerful flowers. They grow well in full shade but like to have at least a little sunlight. A location that gets sun in early spring then light shade through summer is ideal. Wood poppy is native to the eastern U.S. and Ontario. In Ontario, only a couple of small stands remain, and wood poppy is considered an endangered species in Canada. Do not dig plants from the wild. Plants grow easily from seed and will self-seed and spread in good growing conditions. Unwanted seedlings are easy to pull up.

Planting

Seeding: sow seeds directly or in containers in a cold frame in fall
Transplanting: spring
Spacing: 30 cm

Growing

Wood poppy grows best in **partial to full shade** in **moist, humus-rich** soil of **moderate fertility** but tolerates a range of soil conditions as long as the soil remains moist. This plant will go dormant if it does not receive adequate moisture but will return in spring. Topdress with leaf mould or good, fungally dominant compost. Keep wood poppy out of the hot sun. Deadhead to encourage more blooming and to minimize self-seeding. Divide the thick rhizomes of mature plants in spring.

Tips

Wood poppy is a perfect plant for naturalizing in a woodland garden and does well under many different tree species. It will also do well in beds, borders and wildflower gardens. Unlike many members of the poppy family, wood poppy transplants easily and can be shared with friends.

Recommended

S. diphyllum is an upright plant that spreads by thick rhizomes and has large, basal, deeply lobed, green to blue-green foliage with silvery undersides. Bright yellow to yellow-orange flowers are borne singly or in small clusters in spring, with intermittent blooming throughout summer. Pendent, ovoid, hairy seedpods follow the flowers.

Problems & Pests

Wood poppy is relatively pest free because of the toxic foliage, but may experience damage from slugs and snails. The toxic foliage also discourages consumption by deer and other browsers.

S. diphyllum

Woodrush

Luzula

Height: 30–80 cm • **Spread:** 30–45 cm • **Flower colour:** white, brown • **Blooms:** mid-spring to early summer • **Hardiness:** zones 4–8

Woodrushes are definitely underused plants. These tough, semi-evergreen to evergreen perennials can handle most of what Mother Nature throws at them and still look great. They thrive in damp, shady locations and spread relatively slowly by stolons. They will also self-seed in favourable conditions but are easy to keep under control.

Planting

Seeding: sow seeds directly or in containers in a cold frame in fall or early spring.
Transplanting: spring
Spacing: 30–45 cm

Growing

Woodrushes grow best in **full to partial shade**, and they tolerate full sun if they have consistently moist soil. The ideal soil should be of **average fertility, humus rich, neutral to acidic, moist** and **well drained**, but plants will adapt to a range of soils including clay. *L. sylvatica* is tolerant of dry shade conditions. Plants exposed in winter may suffer burn damage to the foliage. Cut or trim away dead or damaged foliage in spring. Divide in spring or fall to propagate plants or to boost vigour in plants with diminished growth.

Tips

Woodrushes are great for naturalizing in a woodland garden and are wonderful when massed at the edge or in the middle of beds and borders. They are good weed-suppressing groundcovers. The cut flowers work well in fresh arrangements, and the seed heads are attractive in dried arrangements.

Recommended

L. nivea (snowy woodrush) forms loose clumps of arching, medium to deep green or grey-green, grass-like leaves 30–40 cm tall. It bears clusters, up to 60 cm tall, of white flowers from late spring to early summer. Brown seed heads follow the flowers.

L. nivea

L. sylvatica (greater woodrush) forms a dense, 20–35 cm tall clump of glossy, wide, bright green to dark green leaves. It bears small clusters of brown flowers in mid-spring to early summer on stems up to 70 cm tall. **'Aurea'** has bright yellow-green foliage in spring that turns darker in summer and lighter again in fall. **'Marginata'** has narrow, creamy white margins. The cultivars may only be hardy to zone 5.

Problems & Pests

Woodrushes may occasionally experience leaf spot and rust.

L. sylvatica

Yellow Wax-Bells

Kirengeshoma

Height: 60 cm–1.2 m • **Spread:** 60–90 cm • **Flower colour:** yellow • **Blooms:** late summer to early fall • **Hardiness:** zones 5–8

These beautiful, bushy, relatively low-maintenance plants have excellent foliage and wonderful, late-season flowers that feel good when touched. Yellow wax-bells are native to wooded and mountainous areas in Japan and Korea and are part of the Hydrangea family. These are choice plants for shaded gardens, and savvy gardeners snap them up in a hurry. If you see any in the nursery, buy some and plant them in as close to the ideal conditions as possible. Provide enough moisture for the best performance. Plants may take a few years to reach a good size but are well worth the wait.

Planting

Seeding: sow seeds in containers in a cold frame in fall or early spring; germination times vary; cold stratification may improve germination.
Transplanting: spring
Spacing: 60–90 cm

Growing

Yellow wax-bells grow best in **light to partial shade**, but tolerate deeper shade. The soil should be **deep, fertile, well-drained, humus rich, acidic** and **moist**. Plants perform poorly in dry or overly wet soil. Topdress annually with well-composted leaf mould. These plants rarely need dividing, but it can be done carefully in spring or fall. Plants may benefit from shelter from the wind. Cut plants to the ground after they have gone dormant.

Tips

Yellow wax-bells looks great as specimens and when mass planted. They thrive in moist woodland areas, where they should be allowed to spread and form colonies. They can be used at the back of beds and borders, next to ponds and other water features and in containers. The fresh flowers are good for cutting.

Recommended

K. palmata is a shrubby perennial that spreads slowly by thick rhizomes and has upright to arching purplish stems with pointy-lobed, medium to dark green, horizontally oriented and layered foliage. Leaves can reach 20 cm in width. It bears loose clusters of pendent, waxy,

K. palmata (both photos)

bell-shaped, pale yellow flowers with thick fleshy petals at the tops of the stems, followed by interesting, three-pronged, brownish seed heads. The rounded, yellow flower buds swell to a good size before the flowers open.

Problems & Pests

Yellow wax-bells have no serious pest problems other than the odd occurrence of slugs and snails on new growth.

HEIGHT LEGEND: Low: < 30 cm • Medium: 30–60 cm • Tall: > 60 cm

SPECIES
by Common Name

SPECIES by Common Name	COLOUR									BLOOMING			HEIGHT		
	White	Pink	Red/Brown	Orange	Yellow	Blue	Purple	Green	Foliage	Spring	Summer	Fall	Low	Medium	Tall
Anise-Hyssop	•						•				•				•
Arum	•							•			•		•		
Bear's Breeches	•						•			•	•				•
Blackberry Lily				•	•						•				•
Bloodroot	•									•			•		
Bloody Dock			•					•	•		•		•	•	
Blue Bugloss						•				•	•			•	
Blue-Eyed Grass					•	•	•			•	•		•	•	
Bluestar	•					•				•				•	•
Boltonia	•	•					•				•	•			
Bowman's Root	•	•								•	•				•
Brunnera						•				•				•	
Burnet	•		•				•				•	•			•
Buttercup	•				•						•	•	•		•
Calamint	•	•					•				•	•		•	
Carolina Lupine					•					•	•			•	•
Cortusa	•	•					•			•	•		•		
Crocosmia				•	•	•					•			•	
Culver's Root	•	•					•				•	•			•
Cyclamen	•	•	•							•			•		
Daffodil	•	•		•	•					•			•	•	
Edelweiss	•				•						•		•		
Fairy Bells	•				•					•	•		•	•	
False Indigo	•						•			•	•				•
Foamy Bells	•	•							•	•	•		•	•	
Fringe Cups	•	•						•		•	•				•
Gas Plant	•	•					•				•			•	•
Germander		•					•			•	•		•	•	

LIGHT				SOIL CONDITIONS									SPECIES by Common Name
Sun	Part Sahde	Light Shade	Shade	Moist	Well Drained	Dry	Humus Rich	Fertile	Average	Poor	Hardiness Zones	Page Number	
•					•						3–8	54	Anise-Hyssop
•	•			•			•	•			5–8	58	Arum
•	•	•	•		•						5–8	60	Bear's Breeches
•	•			•	•		•		•		5–8	62	Blackberry Lily
	•	•	•	•	•		•	•			3–8	64	Bloodroot
•	•			•	•				•		4–8	66	Bloody Dock
•				•	•				•		3–8	68	Blue Bugloss
•				•	•				•	•	3–8	70	Blue-Eyed Grass
•	•				•				•		3–8	72	Bluestar
•				•	•		•	•			4–8	74	Boltonia
	•	•		•	•		•				4–8	76	Bowman's Root
			•	•	•		•		•		3–8	78	Brunnera
•				•	•						2–8	80	Burnet
•	•	•		•	•						3–8	82	Buttercup
•	•			•	•						5–8	84	Calamint
•	•				•			•			3–8	86	Carolina Lupine
	•	•		•	•		•	•	•		5–8	88	Cortusa
•				•	•			•	•		5–8	90	Crocosmia
•	•			•			•	•	•		3–8	92	Culver's Root
	•	•			•			•	•		5–8	94	Cyclamen
•		•			•				•	•	3–8	96	Daffodil
•					•						3–6	102	Edelweiss
	•	•		•	•		•				4–8	104	Fairy Bells
•					•				•	•	3–8	106	False Indigo
•	•	•		•	•		•	•			3–8	110	Foamy Bells
	•	•		•			•				4–8	112	Fringe Cups
•					•	•		•	•		3–8	114	Gas Plant
•				•	•	•			•		4–8	116	Germander

HEIGHT LEGEND: Low: < 30 cm • Medium: 30–60 cm • Tall: > 60 cm

SPECIES
by Common Name

Species by Common Name	White	Pink	Red/Brown	Orange	Yellow	Blue	Purple	Green	Foliage	Spring	Summer	Fall	Low	Medium	Tall
	COLOUR									BLOOMING			HEIGHT		
Globeflower				•	•					•	•			•	•
Golden Hakone Grass									•				•	•	
Green and Gold					•					•	•		•		
Hair Grass					•		•		•	•	•		•	•	•
Hardy Orchids	•	•								•	•	•	•	•	
Hellebore	•	•			•		•	•		•			•	•	•
Holly Fern									•					•	
Hyssop	•	•					•				•	•			
Ice Plant		•	•	•			•			•	•	•	•		
Indian Pink			•		•					•	•			•	
Inula					•						•	•		•	•
Irish Moss	•									•	•		•		
Ironweed							•				•	•			•
Jack-in-the-Pulpit							•	•		•	•			•	
Jupiter's Beard	•	•	•								•				•
Kalimeris	•					•				•	•	•		•	
Kenilworth Ivy	•						•			•	•	•	•		
Knautia			•								•	•		•	•
Lady's Slipper Orchid	•	•	•		•					•	•		•	•	•
Lemon Balm	•				•						•			•	
Lilyturf	•						•				•	•	•	•	
Liverwort	•	•				•	•			•			•		
Mayapple	•	•	•				•			•	•		•	•	•
Mondo Grass	•						•				•		•		
Mountain Mint	•	•									•	•		•	•
Northern Maidenhair Fern									•		•			•	
Ox-Eye Daisy					•						•	•		•	
Patrinia	•				•						•	•		•	•

Sun	Part Sahde	Light Shade	Shade	Moist	Well Drained	Dry	Humus Rich	Fertile	Average	Poor	Hardiness Zones	Page Number	SPECIES by Common Name
	•			•					•		3–7	118	Globeflower
	•	•		•	•		•			•	5–8	120	Golden Hakone Grass
	•	•	•	•	•		•				5–8	122	Green and Gold
•	•	•		•	•	•	•				3–8	124	Hair Grass
	•			•	•			•	•		5–8	128	Hardy Orchids
		•		•	•			•	•		4–8	130	Hellebore
	•	•	•	•	•			•	•		3–8	134	Holly Fern
•					•				•		4–8	136	Hyssop
•					•	•				•	4–8	138	Ice Plant
	•	•		•	•				•		5–8	140	Indian Pink
•	•			•	•						3–8	142	Inula
•	•	•		•	•				•	•	3–7	144	Irish Moss
•	•			•	•				•		3–8	146	Ironweed
	•	•		•	•			•			4–8	148	Jack-in-the-Pulpit
•					•					•	4–8	152	Jupiter's Beard
•	•			•	•					•	4–8	154	Kalimeris
	•	•			•						4–8	156	Kenilworth Ivy
•					•					•	4–8	158	Knautia
	•	•	•	•	•		•				2–7	160	Lady's Slipper Orchid
•				•	•				•		3–7	162	Lemon Balm
	•	•		•	•			•	•		4–8	164	Lilyturf
	•	•	•		•		•				4–8	168	Liverwort
	•	•	•	•			•	•			5–8	170	Mayapple
	•	•		•			•	•			6–8	174	Mondo Grass
•	•	•		•	•				•		4–8	176	Mountain Mint
	•	•	•	•	•				•		3–8	178	Northern Maidenhair Fern
•				•	•						3–7	180	Ox-Eye Daisy
•	•			•	•			•	•		5–8	182	Patrinia

HEIGHT LEGEND: Low: < 30 cm • Medium: 30–60 cm • Tall: > 60 cm

SPECIES
by Common Name

Species	White	Pink	Red/Brown	Orange	Yellow	Blue	Purple	Green	Foliage	Spring	Summer	Fall	Low	Medium	Tall
Perennial Salvia	•	•				•	•			•	•	•		•	•
Perennial Sunflower					•						•	•			•
Pitcher Plant		•	•		•		•		•	•			•		•
Prairie Coneflower			•		•						•	•			•
Prairie Poppy Mallow			•				•			•	•		•		
Purple Moor Grass			•				•				•	•		•	•
Pussy Toes	•	•								•	•		•		
Ribbon Grass	•	•							•		•				•
Rodgersia	•	•								•	•				•
Rosinweed					•						•	•			•
Sedge			•					•	•	•	•		•	•	•
Self-Heal	•	•					•			•	•	•	•		
Sheep's Bit						•	•				•		•		
Shooting Star	•	•								•			•		
Skullcap	•	•			•	•	•				•	•	•		•
Stokes' Aster	•	•			•	•	•				•	•		•	
Stonecress	•	•								•	•		•		
Toad Lily	•	•					•				•	•	•		•
Trillium	•	•	•		•		•			•				•	
Tuber Oat Grass						•		•	•		•	•	•	•	
Tulip	•	•	•	•	•		•	•		•			•		
Turtlehead	•	•					•				•	•		•	•
Umbrella Plant	•	•								•				•	•
Virginia Bluebells		•				•	•			•					•
Whitlow's Grass					•					•			•		
Wood Fern									•					•	•
Wood Poppy					•					•	•		•		
Woodrush	•		•							•	•			•	•
Yellow Wax-Bells					•						•	•			•

Sun	Part Sahde	Light Shade	Shade	Moist	Well Drained	Dry	Humus Rich	Fertile	Average	Poor	Hardiness Zones	Page Number	SPECIES by Common Name
•					•				•		3–8	184	Perennial Salvia
•				•	•				•		4–8	190	Perennial Sunflower
•	•			•			•				2–8	194	Pitcher Plant
•				•	•	•			•		3–8	196	Prairie Coneflower
•					•	•					4–8	198	Prairie Poppy Mallow
•	•			•				•	•		4–8	200	Purple Moor Grass
•	•					•	•		•	•	1–8	202	Pussy Toes
•	•			•					•		4–8	204	Ribbon Grass
	•	•		•				•	•		3–8	206	Rodgersia
•	•			•					•		5–8	210	Rosinweed
•	•			•	•				•		2–8	212	Sedge
•	•			•	•				•		3–8	216	Self-Heal
•				•	•				•		4–8	218	Sheep's Bit
	•			•	•		•				4–8	220	Shooting Star
•	•	•			•				•	•	4–8	224	Skullcap
•				•	•				•		4–8	226	Stokes' Aster
•					•				•		4–8	230	Stonecress
	•	•		•	•		•	•			4–8	232	Toad Lily
	•	•	•	•	•		•				4–7	236	Trillium
•	•			•	•				•		3–8	240	Tuber Oat Grass
•					•				•		3–8	242	Tulip
•	•			•	•		•	•			3–8	246	Turtlehead
•	•			•					•	•	5–8	248	Umbrella Plant
	•	•		•	•		•		•		3–8	250	Virginia Bluebells
•					•				•	•	2–8	252	Whitlow's Grass
	•	•		•				•	•		3–8	254	Wood Fern
	•	•	•	•				•	•		4–8	258	Wood Poppy
	•	•	•	•	•			•	•		4–8	260	Woodrush
	•	•		•	•			•	•		5–8	262	Yellow Wax-Bells

Pests, Diseases and What to Do

Anthracnose

Fungus. Yellow or brown spots on leaves; sunken lesions and blisters on stems; can kill plant.

What to Do: Choose resistant plants; keep soil well drained; thin out stems to improve air circulation; avoid handling wet foliage. Remove and destroy infected plant parts; clean up and destroy debris from infected plants at end of growing season. Liquid copper spray can prevent spread to other susceptible plants.

Aphids

Tiny, pear-shaped insects, winged or wingless; green, black, brown, red or grey. Cluster along stems, on buds and on leaves. Suck sap from plants; cause distorted or stunted growth.

Aphids

Sticky honeydew forms on surfaces and encourages sooty mould growth.

What to Do: Squish small colonies by hand; dislodge them with water spray; spray serious infestations with insecticidal soap, horticultural oil or neem oil; encourage predatory insects and birds that feed on aphids.

Aster Yellows
see Viruses

Beetles

Many types and sizes; usually rounded in shape with hard, shell-like outer wings covering membranous inner wings. Some are beneficial, e.g., ladybird beetles ("ladybugs"). Others, e.g., Japanese beetles, flea beetles, blister beetles, leaf skeletonizers and weevils, eat plants. Larvae: see Borers, Grubs. Leave wide range of chewing damage: make small or large holes in or around margins of leaves; consume entire leaves or areas between leaf veins ("skeletonize"); may also chew holes in flowers.

Ladybug

What to Do: Pick beetles off at night and drop them into an old coffee can half filled with soapy water (soap prevents them from floating); spread an old sheet under plants and shake off beetles to collect and dispose of them. Use a hand-held vacuum cleaner to remove them from plant. Parasitic nematodes are effective if the beetle goes through part of its growing cycle in the ground.

Japanese beetle

Foliage chewed by beetles

Beneficial predatory ladybird beetle larva

Predatory ground beetle

Blight

Fungal or bacterial diseases, many types; e.g., leaf blight, snow blight, tip blight. Leaves, stems and flowers blacken, rot and die.

What to Do: Thin stems to improve air circulation; keep mulch away from base of plants; remove debris from garden at end of growing season. Remove and destroy infected plant parts.

Borers

Larvae of some moths, wasps, beetles; very damaging plant pests. Worm-like; vary in size and get bigger as they bore through plants. Burrow into stems, leaves and/or roots and rhizomes, destroying vascular tissue and weakening stems to cause breakage. Leaves will wilt; may see tunnels in leaves, stems or roots; rhizomes may be hollowed out entirely or in part.

What to Do: May be able to squish borers within leaves. Remove and destroy bored parts; may need to dig up and destroy infected roots and rhizomes.

Budworms

Moth larvae 13–19 mm long; striped; green, yellow-green, tan, dull red. Bore into buds, eat from inside out; may also eat open flowers and new leaf growth. Buds and new leaves appear tattered or riddled with holes.

What to Do: Pick off by hand daily and drop into a can of soapy water. Remove infested plants and destroy. Don't replant susceptible varieties.

Bugs (True Bugs)

Small insects, up to 13 mm long; green, brown, black or brightly coloured and patterned. Many beneficial; a few pests, such as lace bugs, pierce plants to suck out sap. Toxins may be injected that deform plants; sunken areas left where pierced; leaves rip as they grow; leaves, buds and new growth may be dwarfed and deformed.

What to Do: Always properly identify bugs before implementing controls. Remove debris and weeds from around plants in fall to destroy overwintering sites. Pick off by hand and drop into soapy water. Use parasitic nematodes if part of bug's growth cycle is in the ground. Spray plants with insecticidal soap or neem oil.

Spittlebugs

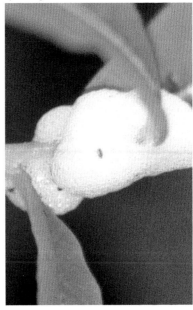

Canker

Swollen or sunken lesions, often on stems, caused by many different bacterial and fungal diseases. Most canker-causing diseases enter through wounds.

What to Do: Maintain plant vigour; avoid causing wounds; control borers and other tissue-dwelling pests. Prune out and destroy infected plant parts. Sterilize pruning tools before and after use.

Caterpillars

Larvae of butterflies, moths, sawflies. Examples: budworms (see Budworms), cutworms (see Cutworms), leaf rollers, leaf tiers, loopers. Chew foliage and buds; can completely defoliate plant if infestation is severe.

What to Do: Removal from plant is best control. Use high-pressure water and soap, or pick caterpillars off by hand. Control biologically using *B.t.*

Cutworms

Larvae of some moths. About 2.5 cm long, plump, smooth skinned; curl up when poked or disturbed. Usually affect young plants and seedlings, which may be completely consumed or chewed off at ground level.

What to Do: Pick off by hand. Use old toilet tissue rolls to make barrier collars around plant bases; push tubes at least halfway into the ground.

Damping Off
see p. 39

Galls

Unusual swellings of plant tissues that may be caused by insects, such as Hemerocallis gall midge, or by diseases. Can affect leaves, buds, stems, flowers, fruit. Often a specific gall affects a single genus or species.

What to Do: Cut galls out of plant and destroy them. Galls caused by

Caterpillar

Remove and destroy any infected plant parts. Use horticultural oil as a preventive measure. Compost tea is also effective.

Grubs

Larvae of different beetles, commonly found below soil level; usually curled in C-shape. Body white or grey; head may be white, grey, brown or reddish. Problematic in lawns; may feed on plant roots. Plant wilts despite regular watering; may pull easily out of ground in severe cases.

What to Do: Toss any grubs found while digging onto a stone path or patio for birds to devour; apply parasitic nematodes.

Leaf Blotch
see Leaf Spot

Leafhoppers & Treehoppers

Small, wedge-shaped insects; can be green, brown, grey or multi-coloured. Jump around frantically when disturbed. Suck juice from plant leaves, cause distorted growth, carry diseases such as aster yellows.

What to Do: Encourage predators by planting nectar-producing plants. Wash insects off with strong spray of water; spray with insecticidal soap or neem oil.

Leaf Miners

Tiny, stubby larvae of some butterflies and moths; may be yellow or green. Tunnel within leaves leaving winding trails; tunnelled areas lighter

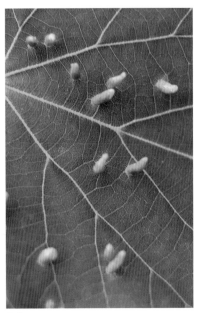

Galls

insects usually contain the insect's eggs and juvenile stages. Prevent such galls by controlling insects before they lay eggs; otherwise, try to remove and destroy infected tissue before young insects emerge. Insect galls generally more unsightly than damaging to plant. Galls caused by diseases often require destruction of plant. Do not place other plants susceptible to same disease in that location.

Grey Mould (Botrytis Blight)

Fungal disease. Leaves, stems and flowers blacken, rot and die.

What to Do: Thin stems to improve air circulation; keep mulch away from base of plant, particularly in spring when plant starts to sprout; remove debris from garden at end of growing season; do not overwater.

Leaf Miners

Sterilize removal tools; avoid wetting foliage or touching wet foliage; remove and destroy debris at end of growing season. Spray plant with liquid copper. Compost tea or a mixture of baking soda and citrus oil also works in most instances.

Mealybugs

Tiny crawling insects related to aphids; appear to be covered with white fuzz or flour. More often found on houseplants than in the garden. Sucking damage stunts and stresses plant. Mealybugs excrete honeydew, promoting sooty mould.

in colour than rest of leaf. Unsightly rather than health risk to plant.

What to Do: Remove debris from area in fall to destroy overwintering sites; attract parasitic wasps with nectar plants. Remove and destroy infected foliage; can sometimes squish by hand within leaf. Floating row covers prevent eggs from being laid on plant. Bright blue sticky cards, available in most nurseries, attract and trap adult leaf miners.

What to Do: Remove by hand from smaller plants; wash off plant with soap and water; wipe off with alcohol-soaked swabs; remove heavily infested leaves; encourage or introduce natural predators such as mealybug destroyer beetle and parasitic wasps; spray with insecticidal soap. Note: larvae of mealybug destroyer beetle look like very large mealybugs.

Leaf Spot

Two common types: one caused by bacteria and the other by fungi. Bacterial: small brown or purple speckles grow to encompass entire leaves; leaves may drop. Fungal: black, brown or yellow spots; leaves wither; e.g., scab, tar spot, leaf blotch.

What to Do: Bacterial infection more severe; must remove entire plant. For fungal infection, remove and destroy infected plant parts.

Mealybugs

Control powdery mildew by spraying foliage with mixture of horticultural oil and baking soda in water. Three applications one week apart needed.

Mites

Tiny, eight-legged relatives of spiders. Examples: spider mites, rust mites, thread-footed mites. Invisible or nearly invisible to naked eye; red, yellow, green or translucent; usually found on undersides of plant leaves. Suck juice out of leaves; may see their fine webs on leaves and stems; may see mites moving on leaf undersides; leaves become discoloured and speckled in appearance, then turn brown and shrivel up.

What to Do: Wash off with a strong spray of water daily until all signs of infestation are gone; predatory mites are available through garden centres; apply insecticidal soap, horticultural oil or neem oil.

Nematodes

Tiny worms that give plants disease symptoms. One type infects foliage and stems; the other infects roots. Foliar: yellow spots that turn brown on leaves; leaves shrivel and wither; problem starts on lowest leaves and works up plant. Root-knot: plant is stunted; may wilt; yellow spots on leaves; roots have tiny bumps or knots.

What to Do: Mulch soil, add organic matter, clean up debris in fall; don't touch wet foliage of infected plants. Can add parasitic nematodes to soil. Remove infected plants in extreme cases.

Powdery mildew

Mildew

Two types, both caused by fungus, but with slightly different symptoms. Downy mildew: yellow spots on upper sides of leaves and downy fuzz on undersides; fuzz may be yellow, white or grey. Powdery mildew: white or grey powdery coating on leaf surfaces that doesn't brush off.

What to Do: Choose resistant cultivars; space plants well; thin stems to encourage air circulation; tidy any debris in fall. Remove and destroy infected leaves or other parts. Spray compost tea or highly diluted fish emulsion (1 tsp. per qt. of water) to control downy and powdery mildew.

Rot

Several different fungi or bacteria that affect different parts of a plant and can kill it. Bacterial soft rot: enters through wounds; begins as small, water-soaked lesions on roots and leaves. As lesions grow, their surfaces darken but remain unbroken, while underlying tissue becomes soft and mushy. Lesions may ooze if surface broken. Black rot: bacterial; enters through pores or small wounds. Begins as V-shaped lesions along leaf margins. Leaf veins turn black and eventually plant dies. Crown rot (stem rot): fungal; affects base of plant, causing stems to blacken and fall over and leaves to yellow and wilt. Root rot: fungal; leaves yellow and plant wilts; digging up plant shows roots rotted away.

Fungal disease

What to Do: Keep soil well drained; don't damage plant when digging around it; keep mulches away from plant base. Remove infected plants.

Rust

Fungi. Pale spots on upper leaf surfaces; orange, fuzzy or dusty spots on leaf undersides. Examples: blister rust, hollyhock rust, white rust.

What to Do: Choose varieties and cultivars resistant to rust; avoid handling wet leaves; provide plant with good air circulation; use horticultural oil to protect new foliage; clean up garden debris at end of growing season. Remove and destroy infected plant parts; do not put infected plants in compost pile.

Scale Insects

Tiny, shelled insects that suck sap, weakening and possibly killing plant or making it vulnerable to other problems. Scale appears as tiny bumps typically along stems or on undersides of foliage. Once female scale insect has pierced plant with mouthpart, it is there for life. Juvenile scale insects are called crawlers.

What to Do: Wipe off with alcohol-soaked swabs; spray with water to dislodge crawlers; prune out heavily infested branches; encourage natural predators and parasites; spray horticultural oil in spring before bud break.

Slugs & Snails

Both mollusks. Snails have a spiral shell, slugs have no shell; both have slimy, smooth skin. Can be up to 20 cm long; grey, green, black, beige, yellow or spotted. Leave large ragged holes in leaves and silvery slime trails on and around plants.

Snail

What to Do: Remove slug habitat, including garden debris or mulches around plant bases. Use slug-repellent mulches (see sidebar). Increase air circulation. Pick off by hand in the evening and squish with boot or drop in can of soapy water. Spread diatomaceous earth (available in garden centres; do not use the kind meant for swimming pool filters) on soil around plants to pierce and dehydrate the soft slug or snail bodies. Commercial slug and snail baits are effective; some new formulations are not toxic to pets and children. Stale beer in a sunken, shallow dish may be effective. Attach strips of copper to wood around raised beds or to small boards inserted around susceptible groups of plants; slugs and snails get shocked if they touch copper surfaces.

Smut

Fungus that affects any plant parts above ground, including leaves, stems and flowers. Forms fleshy white galls that turn black and powdery.

What to Do: Remove and destroy infected plants. Do not place same plants in that spot for next few years.

Sooty Mould

Fungus. Thin black film forms on leaf surfaces and reduces amount of light getting to leaf surfaces.

What to Do: Wipe mould off leaf surfaces; control insects such as aphids, mealybugs, whiteflies (honeydew left on leaves encourages mould).

Spider Mites
see Mites

Thrips

Tiny insects, difficult to see; may be visible if you disturb them by blowing gently on an infested flower. Yellow, black or brown with narrow, fringed wings. Suck juice out of plant cells, particularly in flowers and buds, causing grey-mottled petals and leaves, dying buds and distorted, stunted growth.

What to Do: Remove and destroy infected plant parts; encourage native predatory insects with nectar plants; spray severe infestations with insecticidal soap or with horticultural oil. Use blue sticky cards to attract and trap adults.

Viruses

Plant may be stunted and leaves and flowers distorted, streaked or discoloured. Examples: aster yellows, mosaic virus, ringspot virus.

Mosaic Virus

What to Do: Viral diseases in plants cannot be treated. Destroy infected plants; control insects such as aphids, leafhoppers and whiteflies that spread disease.

Weevils
see Beetles

Whiteflies

Tiny, white, moth-like insects that flutter up into the air when the plant is disturbed. Live on undersides of leaves and suck juice out, causing yellowed leaves and weakened plants; leave sticky honeydew on leaves, encouraging sooty mould.

What to Do: Usual and most effective remedy is to remove infested plant so insects don't spread to rest of garden. Destroy weeds where insects may live. Attract native predatory beetles and parasitic wasps with nectar plants. Spray severe cases with insecticidal soap. Use yellow sticky cards or make your own sticky trap: mount tin can on stake, wrap can with yellow paper and cover with small, clear plastic bag smeared with petroleum jelly; replace bag when full of flies. Plant sweet alyssum in immediate area. Make a spray from old coffee grounds.

Wilt

If watering hasn't helped a wilted plant, one of two wilt fungi may be at fault. *Fusarium* wilt: plant wilts, leaves turn yellow then die; symptoms generally appear first on one part of plant before spreading to other parts. *Verticillium* wilt: plant wilts; leaves curl up at edges; leaves turn yellow then drop off; plant may die.

What to Do: Both wilts are difficult to control. Choose resistant plant varieties and cultivars; clean up debris at end of growing season. Destroy infected plants; solarize (sterilize) soil before replanting—contact your local garden centre for assistance.

Worms
see Caterpillars, Nematodes

Acid soil: soil with a pH lower than 7.0

Alkaline soil: soil with a pH higher than 7.0

Basal leaves: leaves that form from the crown

Basal rosette: a ring or rings of leaves growing from the crown of a plant at or near ground level; flowering stems of such plants grow separately from the crown

Crown: the part of a plant where the shoots join the roots, at or just below soil level

Cultivar: a cultivated (bred) plant variety with one or more distinct differences from the parent species, e.g., in flower colour, leaf variegation or disease resistance

Damping off: fungal disease causing seedlings to rot at soil level and topple over

Deadhead: to remove spent flowers to maintain a neat appearance and encourage a longer blooming period

Direct sow: to plant seeds straight into the garden, in the location you want the plants to grow

Disbud: to remove some flower buds to improve the size or quality of the remaining ones

Dormancy: a period of plant inactivity, usually during winter or other unfavourable climatic conditions

Double flower: a flower with an unusually large number of petals, often caused by mutation of the stamens into petals

Genus: category of biological classification between the species and family levels; the first word in a scientific name indicates the genus, e.g., *Digitalis* in *Digitalis purpurea*

Hardy: capable of surviving unfavourable conditions, such as cold weather

Humus: decomposed or decomposing organic material in the soil

Hybrid: a plant resulting from natural or human-induced crossbreeding between varieties, species or genera; the hybrid expresses features of each parent plant

Invasive: able to spread aggressively from the planting site and outcompete other plants

Marginal: a plant that grows in shallow water or in consistently moist soil along the edges of ponds and rivers

Neutral soil: soil with a pH of 7.0

Node: the area on a stem from which a leaf or new shoot grows

Offset: a young plantlet that naturally sprouts around the base of the parent plant in some species

pH: a measure of acidity or alkalinity (the lower the pH, the higher the acidity); the pH of soil influences availability of nutrients for plants

Rhizome: a root-like, usually swollen stem that grows horizontally underground, and from which shoots and true roots emerge

Rootball: the root mass and surrounding soil of a container-grown plant or a plant dug out of the ground

Rosette: see Basal rosette

Self-seeding: reproducing by means of seeds without human assistance, so that new plants constantly replace those that die

Semi-hardy: a plant capable of surviving the climatic conditions of a given region if protected

Semi-double flower: a flower with petals that form two or three rings

Single flower: a flower with a single ring of typically four or five petals

Spadix: a spike of small flowers clustered around a fleshy axis, usually enclosed by a spathe

Spathe: the showy outer hood that encloses the flower cluster of certain plants

Species: the original plant from which a cultivar is derived; the fundamental unit of biological classification, indicated by a two-part scientific name, e.g., *Digitalis purpurea* (*purpurea* is the specific epithet)

Subspecies (subsp.): a naturally occurring, regional form of a species, often isolated from other subspecies but still potentially interfertile with them

Taproot: a root system consisting of one main vertical root with smaller roots branching from it

Tender: incapable of surviving the climatic conditions of a given region; requires protection from frost or cold

True: describes the passing of desirable characteristics from the parent plant to seed-grown offspring; also called breeding true to type

Tuber: a swollen part of a rhizome or root, containing food stores for the plant

Variegation: describes foliage that has more than one colour, often patched or striped or bearing differently coloured leaf margins

Variety (var.): a naturally occurring variant of a species; below the level of subspecies in biological classification; also applied to forms produced in cultivation, which are properly called cultivars

About the Author

Don Williamson has turned his passion for gardening into his life's work. His background is in landscaping, golf course construction and management, and in the design and construction of formal landscape settings. With a degree in Applied Horticultural Technology and professional certificates in Turf Management, he has written and co-written several gardening books.